Fractional Random Vibrations I

This two-volume set provides a comprehensive study of fractional random vibration from the perspective of theory and practice. Volume I deals succinctly with the theories of fractional processes and fractional vibration systems.

A major focus of fractional vibrations is the derivation of analytical expressions for the frequency transfer functions of seven classes of fractional vibrations using elementary functions. This is considered from the perspective of the functional form of linear vibrations with frequency-dependent mass, damping, or stiffness. The present results serve as a basis for the study of the novel and frontier topic of fractional processes passing through fractional vibration systems, which is discussed in Volume II.

The title will be essential reading for students, mathematicians, physicists, and engineers interested in fractional random vibration phenomena.

Ming Li is a professor at Ocean College, Zhejiang University, China, and an emeritus professor at East China Normal University. He has been a contributor for many years to the fields of mathematics, statistics, mechanics, and computer science. His publications with CRC Press also include *Multi-Fractal Traffic and Anomaly Detection in Computer Communications*, *Fractal Teletraffic Modeling and Delay Bounds in Computer Communications*, and *Fractional Vibrations with Applications to Euler-Bernoulli Beams*.

Fractional Order Thinking in Exploring the Frontiers of STEM

Fractional Order Thinking (FOT) is about solving today's complex problems in the physical, social and life sciences using the tools of fractional order calculus (FOC). Very soon after Mandelbrot introduced the fractal paradigm into the scientific lexicon it was shown that the integer order calculus (IOC) could not describe the dynamics of fractal processes. A new kind of calculus was required to construct the equations of motion for fractal dynamic processes which turned out to be fractional order Hamiltonian equations (FOHEs). The FOHEs are one tool in the FOC toolbox which concerns how to apply the operators of differentiation and integration of non-integer orders. Rejecting the fractional calculus is equivalent to saying there are no numbers between neighboring integers.

In this book series, we explore the core motivation of fractional calculus by first showing the "core motivation" of IOC invented by Newton and Leibnitz, the fundamental ideas of which can be traced back to the time of Heraclitus of Ephesus. Our ultimate message is that the "IOC" is driven by "the desire and the need" for the "quantification of changes" based on energy gradients in complex dynamic networks, whereas "FOC" is driven by "the desire and the need of understanding complexity" based on information gradients in those same networks. In science, technology, engineering, and mathematics (STEM), metaphorically speaking, there is plenty of room between the integers to enable the archiving of better than the best modelling, control performance, robustness, resilience, and even intelligence.

There is an increasing interest in fractional order dynamic systems (FODS) and controls in the recent research literature, not only because of their novelty but also due to their practical applications. But to accomplish all of this requires a new way of thinking, the Fractional Order Thinking (FOT) referred to above, which in turn must be preceded by a new STEM curriculum. This series will offer a unique platform to demonstrate such additional benefits in our improved understanding of complexity and stochastic dynamics via FOT.

This exciting book series features:

- A forum demonstrating the good consequences of using fractional order thinking (FOT).

- Fractional calculus concepts are presented in the context of complex systems characterization while assuming minimal background in math and physics.

- A broad audience including professionals across many fields, the general public and courses in colleges and even high schools.

- Wide STEM topics ranging from batteries to medicine with a writing style easy to follow.

- Short and inexpensive books of 120–150 pages main text that can be written and read in a reasonable amount of time.

Please contact the series editors, Bruce J. West (North Carolina State University, USA) and YangQuan Chen (University of California Merced), and Taylor & Francis Publisher Lian Sun (Lian.Sun@informa.com), if you have an idea for a book for the series.

Titles in the series currently include:

Fractional Calculus for Skeptics I
The Fractal Paradigm
Bruce J. West and YangQuan Chen

On the Fractal Language of Medicine
Bruce J. West and W. Alan C. Mutch

Fractional Random Vibrations I
Theories
Ming Li

Fractional Random Vibrations II
Applications
Ming Li

For more information about this book series, please visit https://www.routledge.com/Fractional-Order-Thinking-in-Exploring-the-Frontiers-of-STEM/book-series/FOT4STEM

Fractional Random Vibrations I
Theories

Ming Li

CRC Press
Taylor & Francis Group
Boca Raton London New York

CRC Press is an imprint of the
Taylor & Francis Group, an **informa** business

Designed cover image: © Ming Li

First edition published 2026
by CRC Press
2385 NW Executive Center Drive, Suite 320, Boca Raton FL 33431

and by CRC Press
4 Park Square, Milton Park, Abingdon, Oxon, OX14 4RN

CRC Press is an imprint of Taylor & Francis Group, LLC

© 2026 Ming Li

ISBN: 978-1-041-11019-4 (hbk)
ISBN: 978-1-041-11020-0 (pbk)
ISBN: 978-1-041-11174-0 (set)
ISBN: 978-1-003-65789-7 (ebk)

DOI: 10.1201/9781003657897

Typeset in Minion
by Apex CoVantage, LLC

To my wife Yonglan Zhang, my daughter Joanna Jiayue Li, and my parents, Xidong Li and Fanggui Yin — for making it both possible and worthwhile

Contents

Preface

FRACTIONAL CALCULUS AND ITS applications attract increasing interests of researchers in all fields where calculus is utilized, including mechanics of solids and so on. The theme of this monograph is the theories and applications of fractional vibrations driven by random processes and fractional processes in particular. The work is split into two volumes. Volume I deals with the theories and Volume II with applications.

In Volume I, we discuss the fundamentals of harmonic vibrations in Chapter 1, conventional random processes in Chapter 2, fractional processes in Chapter 3, the input-output relationships for vibration systems driven by random signals in Chapter 4, the vibration systems with frequency-dependent mass or damping or stiffness in Chapter 5, the classification of fractional vibration systems in Chapter 6, and the analytic theory of seven classes of fractional vibration systems in Chapter 7, with a postscript to Volume I in Chapter 8.

In Volume II, we address the analytic expressions of the responses to seven classes of fractional vibration systems driven by fully developed ocean surface waves in Chapter 1, fractional Gaussian noise in Chapter 2, generalized fractional Gaussian noise in Chapter 3, fractional Brownian motion in Chapter 4, fractional Ornstein-Uhlenbeck process in Chapter 5, and von Kármán process for wind fluctuation speed in Chapter 6, with a postscript to Volume II in Chapter 7.

This monograph may be a reference for engineers, mathematicians, physicists and postgraduates. Preferred preliminaries are calculus, university physics, theoretic mechanics, mechanics of materials, linear vibrations, and random processes.

The monograph is dedicated to human beings, world peace, and memory of my graduate supervisor, the academician of the Chinese Academy of Engineering, the Chairman of the 12th International Ship and Offshore

Structure Congress, ex-Director of Science Committee of China Ship Scientific Research Center, ex-vice Director of China Ship Scientific Research Center, Professor, Dr. Bing-Han Xu.

Ming Li, Ph.D., Professor, Grand Secretary Li XVI (Rizhao)
Ocean College, Zhejiang University, PR. China

Preface to Volume I

VOLUME I FEATURES SEVEN CHAPTERS, consisting of the theories of fractional random vibrations. Chapters 1 and 2 brief the basics of conventional vibrations and traditional random processes, respectively. Chapter 3 discusses fractional processes, which are demanded in applications to be addressed in Volume II. Input-output relationships of random vibration systems are described in Chapter 4. The highlights in Chapter 5 are in three aspects. First, we show some cases of structures that are with frequency-dependent mass, frequency-dependent damping, or frequency-dependent stiffness. Second, we present a general form of vibration system with frequency-dependent mass, damping, or stiffness. Finally, we put forward the closed-form analytic expressions about that system with respect to the vibration parameters (equivalent damping ratio, equivalent natural angular frequencies, and equivalent frequency ratio), responses (free, impulse, step), frequency transfer function, logarithmic decrement, and Q factor. Chapter 5 is not associated with the knowledge of fractional vibrations. However, it may pave the way to note the reasonableness of the frequency-dependent elements (mass, damping, and stiffness) in the theory of fractional vibrations addressed in Chapters 6 and 7. In Chapter 6, we classify fractional vibrations into seven classes. Chapter 7 gives the closed-form expressions of equivalent mass, damping, and stiffness, equivalent vibration parameters (damping ratio, natural angular frequencies, frequency ratio), responses (free, impulse, and step), frequency transfer function, equivalent logarithmic decrement, and equivalent Q factor for each of seven classes of fractional vibration systems.

Ming Li, Ph.D., Professor, Grand Secretary Li XVI (Rizhao)
Ocean College, Zhejiang University, P.R. China

Acknowledgements

THANKS GO TO PROF. Jianxing Leng and Ocean College, Zhejiang University, China, for providing me with the professor position to lecture on the course of ship hull vibrations for undergraduates and postgraduates. Without my teaching in Zhejiang University, it would have been impossible for me to arrange my time to write this monograph. Prof. Xuekang Gu, the vice technical director of China Ship Scientific Research Center (CSSRC); Prof. Yousheng Wu (Ex-Director of the CSSRC, the academician of the Chinese Academy of Engineering); and Prof. Jingjian Chen (CSSRC) are acknowledged.

I would like to take the opportunity to acknowledge a group of professors in Tsinghua University (Beijing) for their education in hard times. They are Qiji Yang, Shiliang Xu, Desheng Wang, Deyun Lin, Dingyue Kou, Baoqin Liu, Jingzhao She, Daimao Lin, Jingxian Zou, Minsheng Hua, Xiaoqing Ding, Xiguang Ma, Xueli Qiao, Xuexia Zhang, Shichang Hou, Zhenming Feng, Jiaguang Fang, Jiaqing Li, Bo Lv, Siming Luo, Guoxiang Zhao, Mengtao Wang, and Xiyuan Yan.

Prof. YangQuan Chen (University of California, Merced) and Prof. Bruce J. West (US Army Research Office, University of Rochester) for their instructions, comments, and encouragement on this monograph are particularly appreciated. Prof. Swee Cheng Lim (Multimedia University) is appreciated for his instructions and discussions in fractional processes. I am grateful to Mr. Yong Chen for improving the resolution of some drawings. Thanks forever go to my uncle (Kerui Li) and aunt (Zhijian Fang) for their love and encouragement.

The views and conclusions contained in this book are those of the author and should not be interpreted as representing the official policies, either expressed or implied, of the Chinese government.

CHAPTER 1

Fundamentals of Harmonic Vibrations

THIS CHAPTER DESCRIBES THE fundamentals of conventionally linear harmonic vibrations. The particularities in this chapter are twofold. One is that fractional vibrations are nonlinear in Section 1.7. The other is the concept of vibrations with frequency-dependent elements (mass, damping, and stiffness) mentioned with exercises in Section 1.9.

1.1 THREE ELEMENTARY PARTS OF VIBRATION SYSTEM

1.1.1 Mass and Inertia

Consider a rigid mass of a body. It moves at the acceleration $a = x''$. The unit of a is m·s^{-2}. Let F_1 be

$$F_1 = \frac{d}{dt}\left(m\frac{dx}{dt}\right). \tag{1.1}$$

The unit of F_1 is N. It is called inertia force. When m is a constant in terms of t in (1.1), according to Newton's second law, one has

$$F_1 = ma. \tag{1.2}$$

When m is a function of vibration angular frequency ω in (1.1) but irrelevant of t, we say that m is frequency dependent. In that case, (1.2) is expressed by

$$F_1 = m(\omega)a. \tag{1.3}$$

DOI: 10.1201/9781003657897-1

The quantity m is called the mass of that body. Its unit is kg. It is a measure of the inertia that body moves. Eq. (1.3) implies that $m(\omega)$ is a frequency-dependent mass.

1.1.2 Stiffness and Restoration

Suppose that the body connects with a spring with the stiffness k. The unit of k is N·m^{-1}. Let

$$F_2 = -kx. \tag{1.4}$$

The unit of F_2 is N. Eq. (1.4) implies that the direction of F_2 is in the opposite of F_1. The quantity k measures the restoration of a spring to its equilibrium state. Hence, F_2 means a restoration force. When k is a function of vibration frequency ω but irrelevant of t, k is frequency dependent and (1.4) can be expressed by (1.5) in the form

$$F_2 = -k(\omega)x. \tag{1.5}$$

1.1.3 Damping and Resistance

When a damper is supposed to be viscous, a vibration system produces a force F_3. It is proportional to the velocity x' in the form

$$F_3 = cx'. \tag{1.6}$$

In (1.6), c is called the damping coefficient and F_3 is damping force. Its unit is N·s·m^{-1}. It is a measure of resisting the movement of a rigid body m. A damper dissipates energy. If c is a function of ω but irrelevant of t, c is frequency dependent and (1.6) becomes (1.7) in the form

$$F_3 = c(\omega)x'. \tag{1.7}$$

1.2 MOTION EQUATION

Denote the initial displacement and initial velocity by x_0 and v_0, respectively. Then, a vibration motion equation is written by

$$\begin{cases} mx'' + cx' + kx = p(t), \\ x(0) = x_0, x'(0) = v_0, \end{cases} \tag{1.8}$$

where $p(t)$ is a driven force (Harris [1], Palley et al. [2]).

In vibration engineering, one usually writes the motion equation in (1.8) by

$$x'' + \frac{c}{m}x' + \frac{k}{m}x = \frac{p(t)}{m}. \tag{1.9}$$

Since the unit of the quantity $\sqrt{\frac{k}{m}}$ is rad/s, we use the symbol ω_n to denote it by

$$\omega_n = \sqrt{\frac{k}{m}}. \tag{1.10}$$

Using (1.10), one can write (1.9) by

$$x'' + \frac{c}{m}x' + \omega_n^2 x = \frac{p(t)}{m}. \tag{1.11}$$

Because ω_n is only related to m and k of a vibration system, irrelevant of excitation $p(t)$, it is called the damping-free natural angular frequency (natural frequency for short) of the system. The natural frequency in Hz is given by

$$f_n = \frac{\omega_n}{2\pi} = \frac{1}{2\pi}\sqrt{\frac{k}{m}}. \tag{1.12}$$

From (1.12), one has the damping-free natural vibration period, denoted by T_n, in the form of (1.13)

$$T_n = \frac{1}{f_n} = 2\pi\sqrt{\frac{m}{k}}. \tag{1.13}$$

Denote the characteristic equation of (1.11) by

$$s^2 + \frac{c}{m}s + \omega_n^2 = 0. \tag{1.14}$$

The roots of (1.14) are called the characteristic roots. They are expressed by (1.15)

$$s_{1,2} = \frac{-\frac{c}{m} \pm \sqrt{\left(\frac{c}{m}\right)^2 - 4\omega_n^2}}{2}. \tag{1.15}$$

Let $c = c_c$ such that the characteristic equation only has a single root in the form

$$s_{1,2} = -\frac{c_c}{2m}.\tag{1.16}$$

Eq. (1.16) implies

$$\frac{c_c}{m} - 2\omega_n = 0.\tag{1.17}$$

Taking into account (1.10) and (1.17), we have (1.18) to express c_c in the form

$$c_c = 2m\omega_n = 2\sqrt{mk}.\tag{1.18}$$

Define a parameter ζ by (1.19)

$$\varsigma = \frac{c}{c_c} = \frac{c}{2\sqrt{mk}} = \frac{c}{2m\omega_n}.\tag{1.19}$$

It is termed damping ratio. It is a dimensionless parameter. As

$$x'' + \frac{c}{m}x' + \omega_n^2 x = x'' + \varsigma\frac{c_c}{m}x' + \omega_n^2 x,$$

one writes the motion equation by

$$x'' + 2\varsigma\omega_n x' + \omega_n^2 x = \frac{p(t)}{m}.$$

Considering the initial conditions, we have the conventional expression of motion equation in the form

$$\begin{cases} x'' + 2\varsigma\omega_n x' + \omega_n^2 x = \dfrac{p(t)}{m}, \\ x(0) = x_0, x'(0) = v_0. \end{cases}\tag{1.20}$$

1.3 FREE VIBRATION AND DAMPING CLASSIFICATION

If $p(t) = 0$, the system (1.20) suffers from free vibration. The solution of (1.20) is called free response when $p(t) = 0$. A free response is only excited by initial conditions.

Since the characteristic equation of (1.20) is now written by

$$s^2 + 2\varsigma\omega_n s + \omega_n^2 = 0,\tag{1.21}$$

one writes the characteristic roots by

$$s_{1,2} = -\varsigma\omega_n \pm \omega_n\sqrt{\varsigma^2 - 1}. \tag{1.22}$$

Based on (1.21) and (1.22), we discuss four types of damping ratios as follows.

1.3.1 Excessive Damping

When $c > c_c$ or $\varsigma > 1$, a vibration system is said to be overdamped or excessively damped. When $\varsigma > 1$, from (1.22), we see that two characteristic roots are real. The free response is therefore given by

$$x(t) = A_1 e^{s_1 t} + A_2 e^{s_2 t}. \tag{1.23}$$

In (1.23), A_1 and A_2 are determined by initial conditions in the form of (1.24)

$$\begin{cases} A_1 = \dfrac{x_0 s_2 - v_0}{s_2 - s_1} = \dfrac{x_0\omega_n\left(\varsigma + \sqrt{\varsigma^2 - 1}\right) + v_0}{2\omega_n\sqrt{\varsigma^2 - 1}}, \\[4mm] A_2 = \dfrac{v_0 - x_0 s_1}{s_2 - s_1} = \dfrac{v_0 - x_0\omega_n\left(-\varsigma + \sqrt{\varsigma^2 - 1}\right)}{-2\omega_n\sqrt{\varsigma^2 - 1}}. \end{cases} \tag{1.24}$$

The free response of a vibration system when $\varsigma > 1$ is given by

$$\begin{aligned} x(t) = & \frac{x_0\omega_n\left(\varsigma + \sqrt{\varsigma^2 - 1}\right) + v_0}{2\omega_n\sqrt{\varsigma^2 - 1}} e^{-\left(\varsigma - \sqrt{\varsigma^2 - 1}\right)\omega_n t} \\[3mm] & - \frac{v_0 - x_0\omega_n\left(-\varsigma + \sqrt{\varsigma^2 - 1}\right)}{2\omega_n\sqrt{\varsigma^2 - 1}} e^{-\left(\varsigma + \sqrt{\varsigma^2 - 1}\right)\omega_n t}. \end{aligned} \tag{1.25}$$

Eq. (1.25) exhibits that $x(t)$ is monotonically decreasing without vibrations for $\varsigma > 1$, see Figure 1.1. From a view of energy, we see that the initial energy provided by x_0 and v_0 is quickly dissipated when $\varsigma > 1$. In general, monotonically decreasing response is trivial in vibrations.

1.3.2 Small Damping

When $0 < \varsigma < 1$ or $0 < c < c_c$, we say that a vibration system is with small damping. In this case, (1.26) gives two characteristic roots in the form

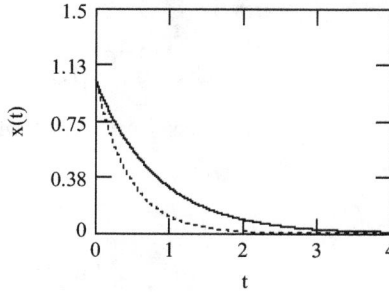

FIGURE 1.1 Overdamped free response $x(t)$ for $f_n = 1$ (Hz), $x_0 = 1$ (m), $v_0 = 1$ (m/s), and $\zeta = 1.2$ (solid), 2.2 (dot).

$$s_{1,2} = -\varsigma w_n \pm w_n\sqrt{\varsigma^2 - 1} = -\varsigma w_n \pm i w_n\sqrt{1 - \varsigma^2}. \qquad (1.26)$$

In (1.26), $i = \sqrt{-1}$. Denote by

$$w_d = w_n\sqrt{1 - \varsigma^2}. \qquad (1.27)$$

The quantity w_d in (1.27) is called damped natural angular frequency (damped natural frequency in short). Let T_d be the period of a damped vibration for $|\zeta| < 1$. Then, it is expressed by (1.28) in the form

$$T_d = \frac{2\pi}{w_d} = \frac{2\pi}{w_n\sqrt{1 - \varsigma^2}}. \qquad (1.28)$$

From (1.10), (1.13), (1.27), and (1.28), one has two inequalities expressed by (1.29)

$$w_d < w_n,$$

$$T_n < T_d. \qquad (1.29)$$

However, w_d makes sense if $|\zeta| < 1$. Since $|\zeta| < 1$ implies $\varsigma^2 \ll 1$, $\varsigma^2 \approx 0$. In fact, there are uncertainties and errors in measurement, structure manufacture, computations, and so on in engineering. Thus, it is reasonable to consider $\sqrt{1 - \varsigma^2} \approx 1$. Therefore, in engineering, we often adopt (1.30)

$$w_d \approx w_n,$$

$$T_n \approx T_d. \qquad (1.30)$$

The free response of a vibration system with small damping is given by

$$x(t) = e^{-\varsigma\omega_n t}\left(x_0\cos\omega_d t + \frac{v_0 + \varsigma\omega_n x_0}{\omega_d}\sin\omega_d t\right), \quad t \geq 0. \tag{1.31}$$

Eq. (1.31) can be rewritten by

$$x(t) = Ae^{-\varsigma\omega_n t}\cos(\omega_d t - \theta), \quad t \geq 0, \tag{1.32}$$

where A is in the form of (1.33)

$$A = \sqrt{x_0^2 + \left(\frac{v_0 + \varsigma\omega_n x_0}{\omega_d}\right)^2}, \tag{1.33}$$

and θ is given by (1.34)

$$\theta = \arctan\frac{v_0 + \varsigma\omega_n x_0}{\omega_d x_0}. \tag{1.34}$$

From the point of view of energy, we see that the initial energy provided by x_0 and v_0 is slowly dissipated by a damper with $0 < \varsigma < 1$. Figure 1.2 indicates the free response of a vibration system with small damping.

1.3.3 Critical Damping

If $c = c_c$ or $\varsigma = 1$, one says that a vibration system is critically damped. When $\varsigma = 1$, one has (1.35) to express the free response in the form

$$x(t) = [x_0 + (v_0 + x_0\omega_n)t]e^{-\varsigma\omega_n t}, \quad t \geq 0. \tag{1.35}$$

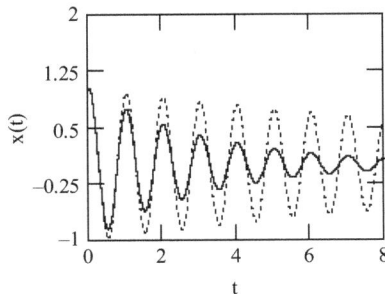

FIGURE 1.2 Small damping-free response $x(t)$ for $f_n = 1$ (Hz), $x_0 = 1$ (m), $v_0 = 1$ (m/s), and $\varsigma = 0.05$ (solid), 0.01 (dot).

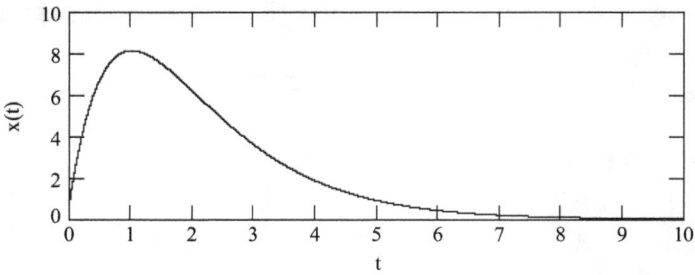

FIGURE 1.3   Critically free response $x(t)$ for $f_n = 3$ (Hz), $x_0 = 1$ (m), $v_0 = 1$ (m/s), and $\zeta = 1$.

Note that a vibration system with critical damping does not oscillate. Critical damping ratio $\zeta = 1$ is a theoretic description. Due to uncertainties and errors in measurement, structure manufacture, computations, and so on, in engineering, a practically critical damping ratio may be described by $\zeta = 1 \pm \varepsilon$, where ε is a positive infinitesimal. When $\zeta = 1 + \varepsilon$, the system slides towards excessively damped case. If $\zeta = 1 - \varepsilon$, the system is towards small damping. Figure 1.3 shows a plot of critically free response $x(t)$.

1.3.4 Negative Damping

If $c < 0$ or $\zeta < 0$, we say that a vibration system is negatively damped. In this case, the free response is expressed by (1.36) in the form

$$x(t) = e^{|\varsigma|\omega_n t}\left(x_0 \cos\omega_d t + \frac{v_0 - |\varsigma|\omega_n x_0}{\omega_d}\sin\omega_d t \right), \quad t \geq 0. \qquad (1.36)$$

From (1.36), one sees that $x(t)$ approaches infinity for $t \to \infty$ until a structure is collapsed or out of control or out of order. Investigating negative damping is crucially meaningful in either academic research or practice, referring to Den Hartog [3], Nakagawa and Ringo [4] for some examples of structures that may cause vibrations with negative damping. Li [5] pays attention to negatively damped fractional vibrations. Figure 1.4 gives a plot of a free response with negative damping.

1.4 FORCED RESPONSES

A forced response is a solution of a motion equation (1.20) for $p(t) \neq 0$. Considering the superposition, one may write $x(t)$ by

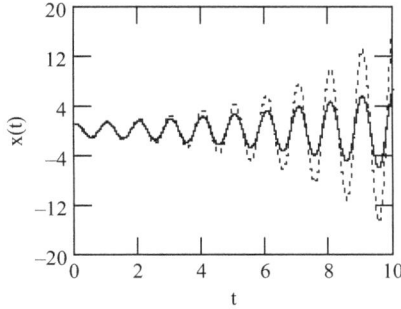

FIGURE 1.4 Free response to a negatively damped vibration system for $f_n = 3$ (Hz), $x_0 = 1$ (m), $v_0 = 1$ (m/s), and $\zeta = -0.010$ (solid), 0.015 (dot).

$$x(t) = x_{zi}(t) + x_{zs}(t). \tag{1.37}$$

In (1.37), $x_{zi}(t)$ is called zero input response and $x_{zs}(t)$ is termed zero state response. The function $x_{zi}(t)$ is the response only excited by the initial conditions x_0 and v_0. The function $x_{zs}(t)$ is the response only driven by $p(t)$ with zero initial conditions (Li [5]). As $x_{zi}(t)$ is the free response previously discussed in Section 1.3, we only discuss $x_{zs}(t)$ in this section and omit the subscript zs without causing confusion in what follows.

1.4.1 Unit Impulse Response
When $p(t) = \delta(t)$ (delta function) and $x_0 = v_0 = 0$, the response is specifically denoted by $h(t)$. It is called unit impulse response (impulse response for short). To be precise, $h(t)$ is the solution to (1.38).

$$\begin{cases} h''(t) + 2\zeta\omega_n h'(t) + \omega_n^2 h(t) = \dfrac{\delta(t)}{m}, \\ h(0) = 0, h'(0) = 0. \end{cases} \tag{1.38}$$

From (1.38), one has

$$h(t) = \frac{1}{m\omega_d} e^{-\zeta\omega_n t} \sin\omega_d t u(t). \tag{1.39}$$

In (1.39), $u(t)$ is the unit step function. Figure 1.5 indicates some plots of $h(t)$ with small damping.

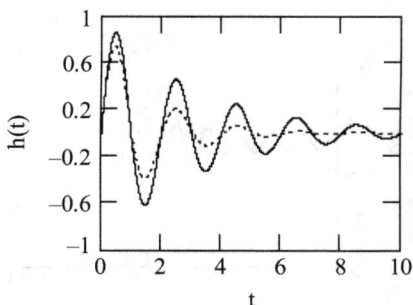

FIGURE 1.5   Impulse response when $\omega_n = \pi$ and $\zeta = 0.1$ (solid), 0.2 (dot).

1.4.2 Unit Step Response

If $p(t) = u(t)$ and $x_0 = v_0 = 0$, the response is specifically written by $g(t)$. It is called unit step response (step response in short). Precisely, $g(t)$ is the solution to (1.40).

$$
\begin{cases}
g''(t) + 2\varsigma\omega_n g'(t) + \omega_n^2 g(t) = \dfrac{u(t)}{m}, \\
g(0) = 0, g'(0) = 0.
\end{cases}
\tag{1.40}
$$

Due to the relationship between $u(t)$ and $\delta(t)$ in the form

$$
u(t) = \int_{0_-}^{t} \delta(t)dt,
$$

$g(t)$ is given by

$$
g(t) = \int_{0_-}^{t} h(t)dt.
\tag{1.41}
$$

From (1.39) and (1.41), one has

$$
g(t) = \frac{1}{k}\left[1 - \frac{e^{-\varsigma\omega_n t}}{\sqrt{1-\varsigma^2}}\cos(\omega_d t - \phi)\right]u(t).
\tag{1.42}
$$

Eq. (1.43) expresses ϕ in (1.42) in the form

$$
\phi = \tan^{-1}\frac{\varsigma}{\sqrt{1-\varsigma^2}}.
\tag{1.43}
$$

Figure 1.6 illustrates some plots of $g(t)$ for $0 < \zeta < 1$.

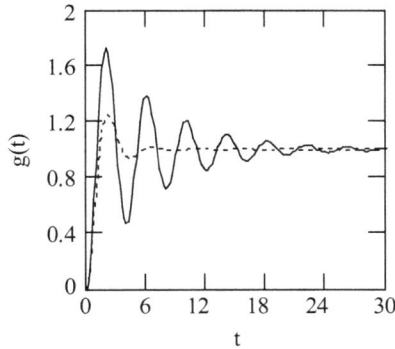

FIGURE 1.6 Step response when $\omega_n = \pi/2$ and $\zeta = 0.2$ (solid), 0.4 (dot).

1.4.3 Sinusoidal Response

When a vibration system is excited by sinusoidal function, the solution to (1.44) is termed the sinusoidal or simple harmonic response.

$$\begin{cases} x'' + 2\zeta\omega_n x' + \omega_n^2 x = \dfrac{A\cos\omega t u(t)}{m}, \\ x(0) = 0, x'(0) = 0. \end{cases} \tag{1.44}$$

According to the convolution theory (Li [5]), we have (1.45).

$$x(t) = h(t) * \frac{A\cos\omega t u(t)}{m} = \frac{1}{m\omega_d} e^{-\zeta\omega_n t} \sin\omega_d t u(t) * \frac{A\cos\omega t u(t)}{m}. \tag{1.45}$$

Doing this convolution yields (1.46) in the form

$$x(t) = \frac{A\cos\omega t}{2m\omega_d} \left\{ \begin{array}{l} \dfrac{e^{-\zeta\omega_n t}[-\zeta\omega_n \sin(\omega_d - \omega)t - (\omega_d - \omega)\cos(\omega_d - \omega)t]}{(\zeta\omega_n)^2 + (\omega_d - \omega)^2} \\[2ex] + \dfrac{e^{-\zeta\omega_n t}[-\zeta\omega_n \sin(\omega_d + \omega)t - (\omega_d + \omega)\cos(\omega_d + \omega)t]}{(\zeta\omega_n)^2 + (\omega_d + \omega)^2} \\[2ex] - \dfrac{(\omega_d - \omega)}{(\zeta\omega_n)^2 + (\omega_d - \omega)^2} - \dfrac{(\omega_d + \omega)}{(\zeta\omega_n)^2 + (\omega_d + \omega)^2} \end{array} \right\}$$

$$+ \frac{A\sin\omega t}{2m\omega_d} \left\{ \begin{array}{l} \dfrac{e^{-\zeta\omega_n t}[(\omega_d - \omega)\sin(\omega_d - \omega)t - \zeta\omega_n \cos(\omega_d - \omega)t]}{(\zeta\omega_n)^2 + (\omega_d - \omega)^2} \\[2ex] + \dfrac{e^{-\zeta\omega_n t}[(\omega_d + \omega)\sin(\omega_d + \omega)t - \zeta\omega_n \cos(\omega_d + \omega)t]}{(\zeta\omega_n)^2 + (\omega_d + \omega)^2} \\[2ex] + \dfrac{\zeta\omega_n}{(\zeta\omega_n)^2 + (\omega_d - \omega)^2} + \dfrac{\zeta\omega_n}{(\zeta\omega_n)^2 + (\omega_d + \omega)^2} \end{array} \right\} \quad t > 0. \tag{1.46}$$

1.5 FREQUENCY TRANSFER FUNCTION

Let $X(\omega)$ be the Fourier transform of $x(t)$. Let $P(\omega)$ be the Fourier transform of $p(t)$. Doing the Fourier transform on both sides of (1.47).

$$x'' + 2\varsigma w_n x' + w_n^2 x = \frac{p(t)}{m} \tag{1.47}$$

produces the motion equation in the frequency domain in the form of (1.48).

$$(i\omega)^2 X(\omega) + 2\varsigma w_n (i\omega) X(\omega) + w_n^2 X(\omega) = \frac{P(\omega)}{m}. \tag{1.48}$$

Note that $X(\omega)$ is the response while $P(\omega)$ is the excitation in the frequency domain. Denote by $H(\omega)$ the frequency transfer function of a vibration system. Considering an excitation $P(\omega)$ that passes through a vibration system, we have (1.49) to express the relationship between $X(\omega)$, $P(\omega)$, and $H(\omega)$ in the form

$$X(\omega) = H(\omega)P(\omega). \tag{1.49}$$

Eq. (1.50) is the expression of $H(\omega)$

$$H(\omega) = \frac{X(\omega)}{P(\omega)}. \tag{1.50}$$

From (1.48), one has

$$H(\omega) = \frac{1}{m\left(w_n^2 - w^2 + i2\varsigma w_n w\right)}. \tag{1.51}$$

Another expression of $H(\omega)$ is written by (1.52).

$$H(\omega) = \frac{1}{mw_n^2\left(1 - \dfrac{w^2}{w_n^2} + i2\varsigma \dfrac{w}{w_n}\right)} = \frac{1}{k\left(1 - \dfrac{w^2}{w_n^2} + i2\varsigma \dfrac{w}{w_n}\right)}. \tag{1.52}$$

Denote by γ the frequency ratio in the form of (1.53).

$$\gamma = \frac{w}{w_n} = \frac{f}{f_n}. \tag{1.53}$$

Using (1.53), one has (1.54) to express $H(\omega)$ by

$$H(\omega) = \frac{1}{k\left(1 - \gamma^2 + i2\varsigma\gamma\right)}. \tag{1.54}$$

In the polar system, one uses (1.55) to express $H(\omega)$ in the form

$$H(\omega) = \left|H(\omega)\right| e^{-i\varphi(\omega)}, \tag{1.55}$$

where $\left|H(\omega)\right|$ and $\phi(\omega)$ are respectively expressed by (1.56) and (1.57).

$$\left|H(\omega)\right| = \frac{1}{k\sqrt{\left(1 - \gamma^2\right)^2 + (2\varsigma\gamma)^2}}, \tag{1.56}$$

$$\varphi(\omega) = \tan^{-1}\frac{2\varsigma\gamma}{1 - \gamma^2}. \tag{1.57}$$

Following Nakagawa and Ringo [4, Chapter 2], (1.58) is considered a convenient expression for drawing curves of $\phi(\omega)$.

$$\varphi(\omega) = \cos^{-1}\frac{1 - \gamma^2}{\sqrt{\left(1 - \gamma^2\right)^2 + (2\varsigma\gamma)^2}}. \tag{1.58}$$

Let F and F^{-1} be the operators of the Fourier transform and its inverse, respectively. Then,

$$H(\omega) = \text{F}[h(t)]. \tag{1.59}$$

Eq. (1.59) implies that (1.60) is true

$$h(t) = \text{F}^{-1}[H(\omega)]. \tag{1.60}$$

1.6 EXCITATION AND RESPONSE RELATIONSHIPS

In the frequency domain, the relationship between excitation $P(\omega)$ and response $H(\omega)$ is expressed by (1.49). Taking into account the convolution rule, one can immediately attain the relationship between $p(t)$ and $x(t)$ in the time domain in the form

$$x(t) = h(t) * p(t). \tag{1.61}$$

In (1.61), $*$ is the convolution operator.

1.7 NOTES ON NONLINEAR VIBRATIONS

The radical requirement in classroom teaching for undergraduates is linear vibrations, as previously discussed. In practice, however, structures may suffer from nonlinear vibrations. There are two categories of nonlinear vibrations. One is conventionally nonlinear vibrations and the other fractional ones.

1.7.1 Conventionally Nonlinear Vibrations

By conventionally nonlinear vibrations, we mean that a motion equation is of integer order but nonlinear. A well-known nonlinear vibration system is described by Duffing's equation in the form

$$mx'' + k(x + \beta x^2) = 0. \tag{1.62}$$

In (1.62), β is a dimensionless coefficient. When $\beta > 0$, we call k is the stiffness of a hard spring. If $\beta < 0$, k is the stiffness of a soft spring. Since the unit of kx is N but $k\beta x^2$ is not, the restoration force in Duffing's equation is non-Newtonian when $\beta \neq 0$.

Another well-known nonlinear vibration system is characterized by the Van de Pol equation in the form

$$mx'' + c|x'|x' + kx = 0. \tag{1.63}$$

The unit of cx' is N but $c|x'|x'$ is non-Newtonian in the Van de Pol equation (1.63).

1.7.2 Fractional Vibrations

Fractional vibrations attract interests of researchers, see, for example, Uchaikin [6, 7], Rossikhin and Shitikova [8–11], Rossikhin [12], Li [13, 14], Duan et al. [15–18]. According to Li [19], (1.64) is a general form of fractional vibrations.

$$m\frac{d^\alpha y(t)}{dt^\alpha} + c\frac{d^\beta y(t)}{dt^\beta} + k\frac{d^\lambda y(t)}{dt^\lambda} = f(t), \quad 1 < \alpha < 3,\ 0 < \beta < 2,\ 0 \leq \lambda < 1. \tag{1.64}$$

Note that (1.64) follows the superposition. To be precise, when excitation $f_j(t)$ ($j = 1, 2, \ldots$) produces the response $y_j(t)$, the excitation $\sum f_j(t)$ yields the response $\sum y_j(t)$. An interesting thing about fractional vibrations is that (1.64) is nonlinear in mechanics. As a matter of fact, the forces of inertia, damping and restoration in (1.64) are non-Newtonian when $\alpha \neq 2$, $\beta \neq 1$, and $\lambda \neq 0$. Precisely, the unit of the fractional inertia force $my^{(\alpha)}(t)$ is not Newtonian for $\alpha \neq 2$. Neither are the fractional damping force $cy^{(\beta)}(t)$ if $\beta \neq 1$ and the fractional restoration one $ky^{(\lambda)}(t)$ when $\lambda \neq 0$. Thus, (1.64) belongs to nonlinear vibrations.

1.8 SUMMARY

We have explained the preliminaries of linear vibrations, referring to [1–5, 20] for more basics of linear vibrations and [19] for resonance bandwidth of fractional vibrations.

1.9 EXERCISES

1.1. Define the normalized frequency transfer function by
$H_1(\omega) = \dfrac{H(\omega)}{H(\omega_n)}$. Prove that $|H_1(\omega)| \leq 1$ if $\zeta > \dfrac{\sqrt{2}}{2}$.

1.2. Suppose that the primary mass of a mass-spring system is a function of time t. Denote it by $m(t)$. Write the expression of $\dfrac{d}{dt}\left[m(t)\dfrac{dx}{dt}\right]$, where x is displacement.

1.3. Suppose that the primary mass of a mass-spring system is a function of angular frequency ω. Denote it by $m(\omega)$. Show that
$\dfrac{d}{dt}\left[m(\omega)\dfrac{dx}{dt}\right] = m(\omega)\dfrac{d^2 x}{dt^2}$.

1.4. Suppose that the primary damping of a mass-spring system is a function of angular frequency ω. Denote it by $c(\omega)$. Show that the friction force is given by $c(\omega)\dfrac{dx}{dt}$.

1.5. Suppose that the primary stiffness of a mass-spring system is a function of angular frequency ω. Denote it by $k(\omega)$. Show that the restoration force is given by $k(\omega)x(t)$.

1.6. Let $X(\omega)$ be the Fourier transform of x. Denote by $F(\omega)$ the Fourier transform of $f(t)$. Write the vibration motion equation
$m(\omega)\dfrac{d^2 x(t)}{dt^2} + c(\omega)\dfrac{dx(t)}{dt} + k(\omega)x(t) = f(t)$ in the frequency domain.

1.7. Let $X(\omega)$ be the Fourier transform of x. Denote by $F(\omega)$ the Fourier transform of $f(t)$. Show that the vibration motion equation of
$m\dfrac{d^2 x(t)}{dt^2} + kx(t) = f(t)$ in the frequency domain is expressed by
$\left[k(\omega) - \omega^2 m(\omega)\right]X(\omega) = F(\omega)$.

1.8. Based on the vibration motion equation $\left[k(\omega) - \omega^2 m(\omega)\right] X(\omega) = F(\omega)$ in the frequency domain, show that the damping-free natural angular frequency is given by $\omega_n = \sqrt{\dfrac{k(\omega)}{m(\omega)}}$.

1.9. Write the expression of damping ratio of the vibration system expressed by $m(\omega)\dfrac{d^2 x(t)}{dt^2} + c(\omega)\dfrac{dx(t)}{dt} + k(\omega)x(t) = f(t)$.

1.10. Write the expression of frequency ratio of the system expressed by $m(\omega)\dfrac{d^2 x(t)}{dt^2} + c(\omega)\dfrac{dx(t)}{dt} + k(\omega)x(t) = f(t)$.

1.11. Write the expression of frequency transfer function of the system $m(\omega)\dfrac{d^2 x(t)}{dt^2} + c(\omega)\dfrac{dx(t)}{dt} + k(\omega)x(t) = f(t)$.

REFERENCES

1. C. M. Harris, *Shock and Vibration Handbook*, 5th Ed., McGraw-Hill, New York, 2002.
2. O. M. Palley, Г. B. Bahizov, and E. Я. Voroneysk, *Handbook of Ship Structural Mechanics*, National Defense Industry Publishing House, Beijing, 2002. In Chinese. Translated from Russian by B. H. Xu, X. Xu, and M. Q. Xu.
3. J. P. Den Hartog, *Mechanical Vibrations*, McGraw-Hill, New York, 1956.
4. K. Nakagawa and M. Ringo, *Engineering Vibrations*, Shanghai Science and Technology Publishing House, Shanghai, China, 1981. In Chinese. Translated from Japanese by S. R. Xia.
5. M. Li, *Fractional Vibrations with Applications to Euler-Bernoulli Beams*, CRC Press, Boca Raton, 2023.
6. V. V. Uchaikin, *Fractional Derivatives for Physicists and Engineers*, vol. I, Springer, Berlin, 2013.
7. V. V. Uchaikin, Relaxation processes and fractional differential equations, *International Journal of Theoretical Physics*, vol. 42, no. 1, 2003, 121–134.
8. Y. A. Rossikhin and M. V. Shitikova, Analysis of the viscoelastic rod dynamics via models involving fractional derivatives or operators of two different orders, *The Shock and Vibration Digest*, vol. 36, no. 1, 2004, 3–26.
9. Y. A. Rossikhin and M. V. Shitikova, Application of fractional calculus for dynamic problems of solid mechanics: Novel trends and recent results, *Applied Mechanics Reviews*, vol. 63, no. 1, 2010, 010801.
10. Y. A. Rossikhin and M. V. Shitikova, Application of fractional operators to the analysis of damped vibrations of viscoelastic single-mass systems, *Journal of Sound and Vibration*, vol. 199, no. 4, 1997, 567–586.

11. Y. A. Rossikhin and M. V. Shitikova, Classical beams and plates in a fractional derivative medium, impact response, in: *Encyclopedia of Continuum Mechanics*, Springer, Berlin, 2020, vol. 1, 294–305.

12. Y. A. Rossikhin, Reflections on two parallel ways in progress of fractional calculus in mechanics of solids, *Applied Mechanics Reviews*, vol. 63, no. 1, 2010, 010701.

13. M. Li, *Theory of Fractional Engineering Vibrations*, Walter de Gruyter, Berlin/Boston, 2021.

14. M. Li, Three classes of fractional oscillators, *Symmetry-Basel*, vol. 10, no. 2, 2018 (91 pages).

15. J.-S. Duan, Y.-J. Lan, and M. Li, A comparative study of responses of fractional oscillator to sinusoidal excitation in the Weyl and Caputo senses, *Fractal and Fractional*, vol. 6, no. 12, 2022, 692.

16. J.-S. Duan, M. Li, Y. Wang, and Y.-L. An, Approximate solution of fractional differential equation by quadratic splines, *Fractal and Fractional*, vol. 6, no. 7, 2022, 369.

17. J.-S. Duan, L. Jing, and M. Li, The mixed boundary value problems and Chebyshev collocation method for Caputo-type fractional ordinary differential equations, *Fractal and Fractional*, vol. 6, no. 3, 2022, 148.

18. J.-S. Duan, D.-C. Hu, and M. Li, Comparison of two different analytical forms of response for fractional oscillation equation, *Fractal and Fractional*, vol. 5, no. 4, 2021, 188.

19. M. Li, Analytic theory of seven classes of fractional vibrations based on elementary functions: A tutorial review, *Symmetry*, vol. 16, no. 9, 2024, 1202.

20. C. Lalanne, *Mechanical Vibration and Shock*, 2nd Ed., John Wiley & Sons, Hoboken, 2009.

CHAPTER 2

Random Processes

THIS CHAPTER DISCUSSES THE basics of conventional random processes. It consists of seven sections. The concept of random process
is in Section 2.1. Ergodicity is in Section 2.2. Three stochastic models,
namely, probability density function, correlation function, and power
spectrum density function, are described in Sections 2.3–2.5, respectively.
Two numeric characteristics are explained in Section 2.6. Section 2.7 is
about the statements of problems regarding conventional random processes. Section 2.8 is the summary.

2.1 CONCEPT OF RANDOM PROCESS

Let $x(t)$ be a random function. By random, we mean that one does not
know what the value of $x(t_1)$ is at t for $t_1 > t$. Due to that, as a variable at t,
$x(t)$ is not deterministic but random, where $t \in \mathbf{R}$ (real numbers set). Thus,
$t \in \mathbf{R}$ is an attribute of a random variable $x(t)$ that occurs at t.

An environment in which $x(t)$ happens is called an experiment. Such
an environment may be either natural or artificial. The term "experiment"
is neither a teaching experiment in a teaching laboratory nor a science one
in a scientific research laboratory. It just implies an abstract environment
where $x(t)$ is produced. In other words, a random variable $x(t)$ is a result
of that experiment. However, in the field of random processes, one utilizes
the term "realization" instead of result to describe $x(t)$. A realization $x(t)$
of an experiment is also called a sample function or a sample in short.

There is a set of experiments that may produce a set of realizations.
One says that the nth experiment results in the nth realization or the nth
sample function. It is denoted by $x_n(t)$, where $n \in \mathbf{N}$ (the set of natural

DOI: 10.1201/9781003657897-2

numbers) and $\mathbf{N}$ is the sample set. Thus, $n \in \mathbf{N}$ is another attribute of a random variable $x_n(t)$.

Though $x_n(t)$ is random, one is able to obtain its probability for $x_n(t) \leq a$. Denote by $\mathrm{Prob}[x_n(t) \leq a] \in [0, 1]$ the probability of $x_n(t) \leq a$. Then, the number $\mathrm{Prob}[x_n(t) \leq a]$ is the third attribute of a random variable $x_n(t)$. Any $x_n(t)$ corresponds to a probability $\mathrm{Prob}[x_n(t) \leq a]$. Denote by $\mathbf{P}$ the probability set. Then, by considering the three attributes of $x_n(t)$, we see that $x_n(t) \in [\mathbf{N}, \mathbf{R}, \mathbf{P}]$.

Let $\{x_n(t)\}$ be the set of sample functions. It is called a random process. A random process designates a set of sample functions $\{x_n(t)\}$. The set $\{x_n(t)\}$ is also called an ensemble of sample functions. Each sample function in $\{x_n(t)\}$ obeys the same probability distribution.

Without confusions occurring, one usually omits the brackets and simply calls $x_n(t)$ a random process. Moreover, by omitting the subscript n, one calls a random process $\{x_n(t)\}$ as a random function and just denotes it by $x(t)$ (Yaglom [1]).

Since $\{x_n(t)\}$ generally includes infinite sample functions, its mathematical expectation (expectation for short) at time t, using the spatial average (Doob [2], Kolmogorov [3]), is given by (2.1).

$$\mu_{xx}(t) = \lim_{N \to \infty} \frac{1}{N} \sum_{n=1}^{N} x_n(t). \tag{2.1}$$

Similarly, (2.2) is a formula for its autocorrelation function (ACF)

$$r_{xx}(t, t+\tau) = \lim_{N \to \infty} \frac{1}{N} \sum_{n=1}^{N} x_n(t) x_n(t+\tau). \tag{2.2}$$

Spatial average plays a role in theory. However, it may be inconvenient in computations in engineering. Practically, infinite sample functions $\{x_n(t)\}$ may not be physically available. For instance, when studying ocean waves in structural vibrations, one encounters difficulty in collecting infinite sample functions of ocean waves during the same period of time. Besides, a practical ocean wave function $x(t)$ may not be considered for infinite length of time but $t \in [0, T]$, where T is a finite positive number. As a matter of fact, a practical random function or time series is only of single history with finite length (Bendat and Piersol [4]). Consequently, spatial average-based computations may usually be inconvenient in practical applications. Fortunately, the ergodicity of stochastic processes provides us with the theory to avoid the mentioned difficulty in computations.

2.2 ERGODICITY

If a process $\{x_n(t)\}$ is stationary and ergodic, the spatial average can be replaced by the time average of any sample function of $\{x_n(t)\}$. Usually, a stationary process $\{x_n(t)\}$ implies that it is ergodic (Priestley [5]). In fact, when $\{x_n(t)\}$ is stationary, it is challenging to find a process that is non-ergodic (Li [6]). For that reason, in default, we regard a stationary process as an ergodic one. When $\{x_n(t)\}$ is stationary and ergodic, therefore, with any $x_n(t)$, expectation is given by

$$\mu_{xx} = \lim_{N \to \infty} \frac{1}{N} \sum_{n=1}^{N} x_n(t) = \lim_{T \to \infty} \frac{1}{T} \int_{-\frac{T}{2}}^{\frac{T}{2}} x_n(t)dt. \qquad (2.3)$$

The right side on (2.3) is time average for expectation of $\{x_n(t)\}$. Similarly,

$$r_{xx}(\tau) = \lim_{N \to \infty} \frac{1}{N} \sum_{n=1}^{N} x_n(t)x_n(t+\tau) = \lim_{T \to \infty} \frac{1}{T} \int_{-\frac{T}{2}}^{\frac{T}{2}} x_n(t)x_n(t+\tau)dt. \qquad (2.4)$$

On the right side of (2.4), $x_n(t)$ is an arbitrary sample function in $\{x_n(t)\}$.

As long as $\{x_n(t)\}$ is stationary, any quantity based on spatial average is equivalent to the one based on time average by using arbitrary sample function $x_n(t)$. This greatly facilitates applying the theory of stochastic process to practice. We omit the subscript n of $x_n(t)$ and directly write $\{x_n(t)\}$ by a random function $x(t)$ unless otherwise stated in what follows.

2.3 PROBABILITY DENSITY FUNCTION

Let $p(x)$ be a probability density function (PDF) of a random function $x(t)$. It is a type of statistical model of $x(t)$. It characterizes the occurrence frequency of x with a certain probability p. Denote by $P(X)$ a probability (cumulative) distribution function of $x(t)$. The relationship between p and $P(X)$ is given by

$$P(X) = P(x \leq X) = \int_{-\infty}^{X} p(x)dx. \qquad (2.5)$$

Eq. (2.6) is the general form of the previous equation.

$$P(x) = P(-a < x \le b) = \int_{-a}^{b} p(x)dx. \qquad (2.6)$$

From (2.5), one has another relationship expressed by (2.7) between p and $P(X)$

$$p(x) = \frac{dP(x)}{dx}. \qquad (2.7)$$

Both $P(x)$ and $p(x)$ are probability models of $x(t)$. Either contains the probability information of $x(t)$.

The function $P(x)$ has the following properties.

- $P(x < \infty) = 1$.

- $P(-\infty < x) = 1$.

- $P(-a < x < b) = p(\xi)(b - a)$ for $\xi \in (-a, b)$, according to the Lagrange mean value theorem.

The literature regarding PDFs is rich. We only mention three in this section, referring to Korn and Korn [7], Papoulis and Pillai [8], and Lu [9], for more PDFs.

2.3.1 PDF of Brownian Motion

Let $x(t)$ be the Brownian motion (Bm) with zero mean. Its PDF is given by

$$p(x) = \frac{1}{\sqrt{4\pi At}} \exp\left(-\frac{x^2}{4At}\right). \qquad (2.8)$$

In (2.8), A is a constant. Since the variance of Bm, denoted by $\mathrm{Var}[x(t)]$, equals to $2At$, Bm is non-stationary.

2.3.2 Gaussian Distribution

Let $x(t)$ be a random function with mean μ and variance σ^2. If its PDF follows (2.9)

$$p(x) = \frac{1}{\sqrt{2\pi}\sigma} e^{-\frac{1}{2}\left(\frac{x-\mu}{\sigma}\right)^2}, \qquad (2.9)$$

it is called a Gaussian process. A Gaussian process is specifically termed normal process. Any other distributions except the normal one are called non-normal distributions. Any other processes except the normal processes are called non-normal processes. Normal processes play a role in random processes.

2.3.3 Poisson Distribution

Let x be a discrete random variable. If, for $\xi > 0$ and $x \in \mathbf{N}$, its PDF is in the form of (2.10).

$$p(x) = e^{-\xi} \frac{\xi^x}{x!}, \tag{2.10}$$

$p(x)$ is called the Poisson PDF and x is a Poisson random function. Eq. (2.11) gives the formula for expectation and variance of a Poisson random function.

$$E(x) = \text{Var}(x) = \xi. \tag{2.11}$$

The Poisson PDF is exponentially decayed.

One thing worth noting is that most PDSs in conventional random processes are exponentially decayed or light tailed.

2.4 CORRELATION FUNCTION

We explain four types of correlation functions. They are autocorrelation function (ACF), auto-covariance function (ACF again), cross-correlation function, and cross-covariance function.

2.4.1 ACF

Let $x(t_1)$ and $x(t_2)$ be two values of $x(t)$ at t_1 and t_2, respectively. Denote by $r_{xx}(t_1, t_2)$ the ACF of $x(t)$. It is a type of statistical model of $x(t)$. It is used to measure how a random variable $x(t_1)$ is statistically correlated to another random variable $x(t_2)$. Precisely, it characterizes how $x(t)$ at one time t_1 correlates to $x(t)$ at another time t_2. An ACF is defined by (2.12).

$$r_{xx}(t_1, t_2) = E[x(t_1)x(t_2)]. \tag{2.12}$$

Eq. (2.13) expresses the expectation or mean of $x(t)$

$$E[x(t)] = \mu_{xx}(t). \tag{2.13}$$

This implies that $E[x(t)]$ may be a function of t.

Denote by $\phi_{xx}(t)$ the strength of $x(t)$. It designates the average power of $x(t)$ as expressed in (2.14).

$$\phi_{xx}(t) = E[x^2(t)]. \tag{2.14}$$

From (2.12), one has

$$E[x^2(t)] = r_{xx}(t, t). \tag{2.15}$$

From (2.14) and (2.15), we have (2.16) to express $\phi_{xx}(t)$ by ACF

$$\phi_{xx}(t) = r_{xx}(t, t). \tag{2.16}$$

Eq. (2.17) gives the variance of $x(t)$ in the form

$$Var[x(t)] = E\{x(t) - E[x(t)]\}^2. \tag{2.17}$$

By (2.15), one has (2.18) to relate $Var[x(t)]$ with ACF in the form

$$Var[x(t)] = r_{xx}(t, t) - E\{[x(t)]\}^2. \tag{2.18}$$

This means that $Var[x(t)]$ may also be a function of t.

One thing important in the above is that a Gaussian process can be uniquely determined by its ACF. In the case of $x(t)$ being stationary, $E[x(t)]$, $Var[x(t)]$ and $r_{xx}(t, t)$ are not dependent on time. In that case, by letting $t_1 = t$, $t_2 = t_1 + \tau$, one has

$$r_{xx}(t_1, t_2) = E[x(t)x(t + \tau)] = r_{xx}(\tau) = \lim_{T\to\infty} \frac{1}{T} \int_{-\frac{T}{2}}^{\frac{T}{2}} x(t)x(t + \tau)dt. \tag{2.19}$$

In (2.19), τ is the time lag. Eqs. (2.20)–(2.23) show that expectation, strength, and variance of a stationary random function are all constants.

$$E[x(t)] = \mu_{xx}(t) = \mu_{xx} = \text{constant}, \tag{2.20}$$

$$\phi_{xx}(t) = \phi_{xx} = r_{xx}(0) = \text{constant}, \tag{2.21}$$

$$\text{Var}[x(t)] = r_{xx}(0) - (\mu_{xx})^2 = \text{constant}. \tag{2.22}$$

Eqs. (2.23) and (2.24) give two properties for $r_{xx}(\tau)$.

$$r_{xx}(0) \geq r_{xx}(\tau), \tag{2.23}$$

$$r_{xx}(\tau) = r_{xx}(-\tau). \tag{2.24}$$

2.4.2 Auto-Covariance Function

When taking $x(t) - \mu_{xx}$ as a stationary random function, one uses auto-covariance function. It is defined by (2.25).

$$C_{xx}(\tau) = E\{[x(t) - \mu_{xx}][x(t + \tau) - \mu_{xx}]\}. \tag{2.25}$$

From (2.25), one has (2.26) to express $\text{Var}[x(t)]$ by $C_{xx}(\tau)$ for $\tau = 0$.

$$\text{Var}[x(t)] = C_{xx}(0). \tag{2.26}$$

Similar to $r_{xx}(\tau)$, (2.27) and (2.28) show two properties for $C_{xx}(\tau)$:

$$C_{xx}(0) \geq C_{xx}(\tau), \tag{2.27}$$

$$C_{xx}(\tau) = C_{xx}(-\tau). \tag{2.28}$$

Denote by $\rho_{xx}(\tau)$ the normalized ACF. By normalized, we mean that $\rho_{xx}(0) = 1$. Eq. (2.29) is its expression

$$\rho_{xx}(\tau) = \frac{r_{xx}(\tau)}{r_{xx}(0)} = \frac{C_{xx}(\tau)}{C_{xx}(0)}. \tag{2.29}$$

Consider the ACF of a Poisson random function $x(t)$. The product $x(t)x(t + \tau)$ is given by $x(t)x(t + \tau) = c^2$ if $x(t)$ and $x(t + \tau)$ are of the same sign. Besides, $x(t)x(t + \tau) = -c^2$ when $x(t)$ and $x(t + \tau)$ are of opposite sign. Thus, for $\lambda > 0$,

$$r_{xx}(\tau) = E[x(t)x(t + \tau)] = c^2 \sum_{n=0}^{\infty} (-1)^n P(n) = c^2 e^{-\lambda|\tau|} \sum_{n=1}^{\infty} (-1)^n \frac{(\lambda|\tau|)^n}{n!} \tag{2.30}$$

$$= c^2 e^{-2\lambda|\tau|}.$$

Eq. (2.30) exhibits that the ACF of the Poisson random function is exponentially decayed.

Note that most ACFs in conventional random processes are fast decayed such that

$$\int_0^\infty r_{xx}(\tau)d\tau < 0. \qquad (2.31)$$

A random function that follows (2.31) is said to be of short-range dependence (SRD) or with short memory.

2.4.3 Cross-Correlation Function

Let $x(t)$ and $y(t)$ be two different random functions. Denote by $r_{xy}(\tau)$ the cross-correlation of $x(t)$ and $y(t)$. It is defined by (2.32).

$$r_{xy}(\tau) = E[x(t)y(t + \tau)] = \lim_{T\to\infty} \frac{1}{T} \int_{-\frac{T}{2}}^{\frac{T}{2}} x(t)y(t+\tau)dt. \qquad (2.32)$$

The cross-correlation $r_{xy}(\tau)$ exhibits how a random function $x(t)$ at t correlates to the other random function $y(t)$ at $t + \tau$. Its properties differ from those of ACF. Generally, $r_{xy}(0)$ may not be the maximum of $r_{xy}(\tau)$. The function $r_{xy}(\tau)$ may be positive or negative. When $r_{xy}(\tau) > 0$, we say that $x(t)$ and $y(t + \tau)$ have a positive correlation, implying they have the same statistical persistence. On the other hand, if $r_{xy}(\tau) < 0$, $x(t)$ and $y(t + \tau)$ are negatively correlated, meaning they have the same statistical anti-persistence.

In general, $r_{xy}(\tau) \neq r_{yx}(\tau)$. However,

$$r_{xy}(\tau) = r_{yx}(-\tau). \qquad (2.33)$$

Eq. (2.33) is true because (2.34) is valid.

$$E[y(t)x(t+\tau)] = r_{yx}(\tau) = \lim_{T\to\infty} \frac{1}{T} \int_{-\frac{T}{2}}^{\frac{T}{2}} y(t)x(t+\tau)dt$$

$$= \lim_{T\to\infty} \frac{1}{T} \int_{-\frac{T}{2}}^{\frac{T}{2}} y(t)x(t+\tau)d(t+\tau) \quad \overset{\text{letting } t+\tau=u}{=} \quad \lim_{T\to\infty} \frac{1}{T} \int_{-\frac{T}{2}+\tau}^{\frac{T}{2}+\tau} y(u-\tau)x(u)du$$

$$= r_{xy}(-\tau). \qquad (2.34)$$

The cross-correlation function $r_{xy}(\tau)$ has the bound expressed by the cross-correlation inequality in the form of (2.35).

$$|r_{xy}(\tau)|^2 \le r_{xx}(0)r_{yy}(0). \qquad (2.35)$$

As a matter of fact, using the Cauchy-Schwartz inequality of (2.36).

$$\left[\int_{-\infty}^{\infty} x(t)y(t)dt\right]^2 \le \int_{-\infty}^{\infty} x^2(t)dt \int_{-\infty}^{\infty} y^2(t)dt, \qquad (2.36)$$

we have

$$|r_{xy}(\tau)|^2 = \left[\lim_{T\to\infty}\frac{1}{T}\int_{-\frac{T}{2}}^{\frac{T}{2}} x(t)y(t+\tau)dt\right]^2 \le \lim_{T\to\infty}\frac{1}{T}\int_{-\frac{T}{2}}^{\frac{T}{2}} x^2(t)dt \lim_{T\to\infty}\frac{1}{T}\int_{-\frac{T}{2}}^{\frac{T}{2}} y^2(t+\tau)dt \qquad (2.37)$$
$$= r_{xx}(0)r_{yy}(0).$$

Eq. (2.37) implies that (2.35) is valid. Another cross-correlation function inequality is expressed by

$$|r_{xy}(\tau)| \le \frac{1}{2}\left[r_{xx}(0)+r_{yy}(0)\right]. \qquad (2.38)$$

In fact, for $a > 0$ and $b > 0$, one has $\sqrt{ab} \le \dfrac{a+b}{2}$. Because $r_{xx}(0) \ge 0$ and $r_{yy}(0) \ge 0$, we have (2.38).

2.4.4 Cross-Covariance Function

When considering the cross-correlation between two random functions $[x(t) - \mu_{xx}]$ and $[y(t) - \mu_{yy}]$, we have the cross-covariance function, denoted by $C_{xy}(\tau)$, in the form of (2.39).

$$C_{xy}(\tau) = E\{[x(t) - \mu_{xx}][y(t+\tau) - \mu_{yy}]\} = r_{xy}(\tau) - \mu_{xx}\mu_{yy}. \qquad (2.39)$$

Similar to $r_{xy}(\tau)$, one has the properties of $C_{xy}(\tau)$. They are expressed by (2.40)–(2.43), respectively.

$$C_{xy}(\tau) = C_{yx}(-\tau), \qquad (2.40)$$

$$|C_{xy}(\tau)|^2 \le C_{xx}(0)C_{yy}(0), \qquad (2.41)$$

$$\left|C_{xy}(\tau)\right| \le \frac{C_{xx}(0) + C_{yy}(0)}{2}. \tag{2.42}$$

When $x = y$, $C_{xy}(\tau)$ reduces to $C_{xx}(\tau) = r_{xx}(\tau) - \mu_{xx}\mu xx$. Due to $0 \le \left|r_{xy}(\tau)\right|^2 \le r_{xx}(0)r_{yy}(0)$, one has

$$0 \le \frac{r_{xy}(\tau)}{\sqrt{r_{xx}(0)r_{yy}(0)}} \le 1. \tag{2.43}$$

In (2.43), one specifically denotes by $\rho_{xy}(\tau)$ the normalized cross-correlation function in the form of (2.44).

$$\rho_{xy}(\tau) = \frac{r_{xy}(\tau)}{\sqrt{r_{xx}(0)r_{yy}(0)}}. \tag{2.44}$$

The normalized cross-correlation function is a quantity to measure the similarity between $x(t)$ and $y(t)$ (Li [10, 11], Fu [12]).

2.5 POWER SPECTRUM

We discuss two types of power spectra. One is power auto-spectrum density function (power spectrum density, PSD for short). The other is power cross-spectrum density function (cross PSD).

2.5.1 Power Auto-Spectrum Density Function

Denote by $X(\omega)$ the Fourier transform of a random function $x(t)$. Then, $X(\omega)$ and $x(t)$ have the same information. The difference between the two is that $x(t)$ is in the time domain while $X(\omega)$ in the frequency domain. A considerably important thing is that $\left|X(\omega)\right|^2$ is a deterministic function, called PSD function.

Denote the PSD of $x(t)$ by $S_{xx}(\omega)$. Then, $S_{xx}(\omega) = \left|X(\omega)\right|^2$. It is a statistical model of $x(t)$ in the frequency domain. According to the Wiener-Khinchin's relation (Khinchin [13], Wiener [14]),

$$S_{xx}(\omega) = \int_{-\infty}^{\infty} r_{xx}(t)e^{-i\omega t}dt. \tag{2.45}$$

In (2.45), $i = \sqrt{-1}$.

Since $r_{xx}(\tau)$ is even, $S_{xx}(\omega)$ is even too in terms of ω. This property is expressed by (2.46).

$$S_{xx}(\omega) = S_{xx}(-\omega) \tag{2.46}$$

Due to (2.46), one may express $S_{xx}(\omega)$ by (2.47).

$$S_{xx}(\omega) = \int_{-\infty}^{\infty} r_{xx}(t)\cos\omega t\, dt. \tag{2.47}$$

For a Poisson random function, its PSD is given by (2.48).

$$S_{\text{Poisson}}(\omega) = \int_{-\infty}^{\infty} r_{xx}(t)e^{-i\omega t}\, dt = \int_{-\infty}^{\infty} c^2 e^{-2\lambda|t|}e^{-i\omega t}\, dt = c^2\frac{4\lambda}{4\lambda^2+\omega^2}. \tag{2.48}$$

If one considers the existence of $\int_{-\infty}^{\infty} r_{xx}(t)e^{-i\omega t}\, dt$ in the domain of ordinary functions, (2.31) must be satisfied, meaning that (2.49) has to be satisfied

$$S_{xx}(0) = \int_{-\infty}^{\infty} r_{xx}(t)dt < 0. \tag{2.49}$$

In short, in the domain of conventional random processes, (2.31) satisfies or $S_{xx}(0)$ is convergent. When (2.49) is satisfied, $x(t)$ is said to be of short-range dependence (SRD) or short memory. However, that is not true for fractional processes (Li [6], Li [15], Bassingthwaighte et al. [16]), where $S_{xx}(0) = \infty$ is a key point of processes with long-range dependence (LRD). We shall emphasize this point in the next chapter.

2.5.2 Cross-Power Spectrum Density Function

Let $S_{xy}(\omega)$ be the cross PSD of $x(t)$ and $y(t)$. Eq. (2.50) is the Wiener-Lee's relation (Korn and Korn [7]) to express $S_{xy}(\omega)$.

$$S_{xy}(\omega) = \int_{-\infty}^{\infty} r_{xy}(t)e^{-i\omega t}\, dt. \tag{2.50}$$

Generally, $S_{xy}(\omega)$ is not an even function in terms of ω. It is usually a complex function. However, (2.51) is true.

$$S_{xy}(\omega) = S_{yx}(-\omega), \tag{2.51}$$

as can be seen from (2.52).

$$S_{xy}(\omega) = \int_{-\infty}^{\infty} r_{xy}(t)e^{-i\omega t}dt = \int_{-\infty}^{\infty} r_{yx}(-t)e^{-i\omega t}dt = S_{yx}(-\omega). \quad (2.52)$$

2.6 TWO NUMERIC CHARACTERISTICS

Two numeric characteristics, namely, mean and variance, are important in random processes.

2.6.1 Mean

When a PDF of $x(t)$ is light tailed, its mean $E(x)$ exists. That implies that the following (2.53) exists.

$$E(x) = \int_{-\infty}^{\infty} xp(x)dx. \quad (2.53)$$

$E(x)$ can also be expressed by time average in the form of (2.54).

$$E(x) = \lim_{T \to \infty} \frac{1}{T} \int_{-\frac{T}{2}}^{\frac{T}{2}} x(t)dt. \quad (2.54)$$

$E(x)$ is a statistical model to describe a global behaviour of $x(t)$ (Li [17]).

2.6.2 Variance

When a PDF of $x(t)$ is of light tail, its variance $Var(x)$ exists, implying that the following (2.55) exists.

$$Var(x) = \int_{-\infty}^{\infty} [x - E(x)]^2 p(x)dx. \quad (2.55)$$

Using time average, $Var(x)$ can be expressed by (2.56).

$$Var(x) = \lim_{T \to \infty} \frac{1}{T} \int_{-\frac{T}{2}}^{\frac{T}{2}} [x - E(x)]^2 dt. \quad (2.56)$$

$Var(x)$ is a statistical model to characterize a local behaviour of $x(t)$ (Li [17]).

Though $E(x)$ and $Var(x)$ are numbers, they play a role in describing global and local behaviours of $x(t)$.

2.7 REMARKS ON CONVENTIONAL RANDOM PROCESSES

If $p(x)$ is heavy tailed, $E(x)$ or $Var(x)$ may not exist (Li [15, 17], Bassingthwaighte et al. [16]). For such an $x(t)$, we greatly desire another two numeric characteristics to measure its global and local properties. Hence, comes the theory of fractal time series or fractional processes (Li [15, 17], Mandelbrot [18], Eliazar and Shlesinger [19], West et al. [20], West and Deering [21], Metzler and Klafter [22], Schreiber [23]).

In the present theory of fractional processes passing through fractional vibration systems in this book, we are interested in a type of random processes whose ACFs are slowly decayed such that

$$\int_{-\infty}^{\infty} r_{xx}(\tau)d\tau = \infty. \tag{2.57}$$

A random process with the property described by (2.57) is said to be of long-range dependence (LRD) or long memory (Beran [24]).

When $S_{xx}(0) = \infty$, one writes $S_{xx}(\omega) \sim 1/\omega$ for $\omega \to 0$. In that case, we say that $x(t)$ is an $1/f$ noise (Mandelbrot [25]). As $S_{xx}(\omega) = \int_{-\infty}^{\infty} r_{xx}(t)e^{-i\omega t}dt$, one has the property described by (2.58) for an LRD $x(t)$

$$S_{xx}(0) = \infty. \tag{2.58}$$

Thus, an LRD process is an $1/f$ noise.

If $x(t)$ is of LRD, its PSD may not exist in the domain of ordinary functions due to $S_{xx}(0) = \infty$. It exists in the domain of generalized functions, however. Again, we need the theory of fractional processes to deal with ACF or PSD of LRD processes or $1/f$ noise.

2.8 SUMMARY

If $x(t)$ is stationary, it usually is ergodic. For an ergodic $x(t)$, its statistic quantities computed by spatial average equal to those by time average with any sample function. For a conventional $x(t)$, five statistic models, namely, PDF, ACF, PSD, mean, and variance, exist. For fractional processes, PSD may not exist in the domain of ordinary functions. Nonetheless, PSD of a fractional process can be taken as a generalized function. In addition, mean and or variance of a fractional process may not exist. Thus, we need other numeric characteristics to measure its global and local behaviours.

They are the Hurst parameter and the fractal dimension; see the next chapter.

2.9 EXERCISES

2.1. Let the ACF be $r_{xx}(\tau) = \dfrac{1}{2}\cos \omega_0 \tau$. Find its Fourier transform.

2.2. Let $r_{xx}(\tau) = A^2 e^{-\alpha|\tau|}$ for $\alpha > 0$. Find its Fourier transform.

2.3. Let $r_{xx}(\tau) = \dfrac{1}{1+\tau^2}$. Find its Fourier transform.

2.4. Let $S_{xx}(\omega) = \sqrt{\dfrac{2|\omega|}{\pi}} K_{0.5}(|\omega|)$, where $K_v(|\omega|)$ is the modified Bessel function of the second kind of order v. Find its inverse Fourier transform.

2.5. Let $r_{x'x'}(\tau)$ be the ACF of $\dfrac{dx(t)}{dt}$. Denote by $r_{xx}(\tau)$ the ACF of $x(t)$.

   Prove that $r_{x'x'}(\tau) = -\dfrac{d^2 r_{xx}(\tau)}{d\tau^2}$.

2.6. Let $S_{x'x'}(\omega)$ be the PSD of $\dfrac{dx(t)}{dt}$. Denote by $S_{xx}(\omega)$ the PSD of $x(t)$.

   Prove that $S_{x'x'}(\omega) = -\omega^2 S_{xx}(\omega)$.

2.7. Denote by $S_{xx}(\omega)$ the PSD of $x(t)$. Let $r_{xx}(\tau)$ be the ACF of $x(t)$. Prove that $S_{xx}(0) = \infty$ if $\displaystyle\int_0^\infty r_{xx}(\tau)d\tau = \infty$.

2.8. Prove that $1/f$ noise is of LRD.

REFERENCES

1. A. M. Yaglom, *An Introduction to the Theory of Stationary Random Functions*, Prentice-Hall, London, 1962.
2. J. L. Doob, *Stochastic Processes*, John Wiley & Sons, New York, 1953.
3. A. N. Kolmogorov, *Foundations of the Theory of Probability*, 2nd Ed., Chelsea Publishing Company, New York, 1956.
4. J. S. Bendat and A. G. Piersol, *Random Data: Analysis and Measurement Procedure*, 4th Ed., John Wiley & Sons, New York, 2010.
5. M. B. Priestley, *Spectral Analysis and Time Series*, Academic Press, London/ New York, 1981.
6. M. Li, *Multi-Fractal Traffic and Anomaly Detection in Computer Communications*, CRC Press, Boca Raton, 2022.

7. G. A. Korn and T. M. Korn, *Mathematical Handbook for Scientists and Engineers*, McGraw-Hill, New York, 1961.

8. A. Papoulis and S. U. Pillai, *Probability, Random Variables, and Stochastic Processes*, 3rd Ed., McGraw-Hill, New York, 2002.

9. D. X. Lu, *Stochastic Processes with Applications*, Tsinghua University Press, Beijing, China, 1986. In Chinese.

10. M. Li, An iteration method to adjusting random loading for a laboratory fatigue test, *International Journal of Fatigue*, vol. 27, no. 7, 2005, 783–789.

11. M. Li, An optimal controller of an irregular wave maker, *Applied Mathematical Modelling*, vol. 29, no. 1, 2005, 55–63.

12. K. S. Fu, editor, *Digital Pattern Recognition*, 2nd Ed., Springer, New York, 1980.

13. A. J. Khinchin, *Mathematical Foundation of Statistical Mechanics*, Dover Publications, Inc., New York, 1949.

14. N. Wiener, *Extrapolation, Interpolation, and Smoothing of Stationary Time Series*, The Technology Press of the MIT and John Wiley & Sons, New York, 1964.

15. M. Li, Fractal time series—a tutorial review, *Mathematical Problems in Engineering*, vol. 2010, Article ID 157264, 26 pages, 2010. https://doi.org/10.1155/2010/157264

16. J. B. Bassingthwaighte, Larry S. Liebovitch, and Bruce J. West, *Fractal Physiology*, Oxford University Press, Oxford, 1994.

17. M. Li, *Fractal Teletraffic Modeling and Delay Bounds in Computer Communications*, CRC Press, Boca Raton, 2022.

18. B. B. Mandelbrot, *Gaussian Self-Affinity and Fractals*, Springer, New York, 2001.

19. I. I. Eliazar and M. F. Shlesinger, Fractional motions, *Physics Reports*, vol. 527, no. 2, 2013, 101–129.

20. B. J. West, E. L. Geneston, and P. Grigolini, Maximizing information exchange between complex networks, *Physics Reports*, vol. 468, no. 1–3, 2008, 1–99.

21. B. J. West and W. Deering, Fractal physiology for physicists Lévy statistics, *Physics Reports*, vol. 246, no. 1–2, 1994, 1–100.

22. R. Metzler and J. Klafter, The random walk's guide to anomalous diffusion: A fractional dynamics approach, *Physics Reports*, vol. 339, no. 1, 2000, 1–77.

23. T. Schreiber, Interdisciplinary application of nonlinear time series methods, *Physics Reports*, vol. 308, no. 1, 1999, 1–64.

24. J. Beran, *Statistics for Long-Memory Processes*, Chapman & Hall, New York, 1994.

25. B. B. Mandelbrot, *Multifractals and 1/f Noisaaaaae*, Springer, New York/Berlin, 1998.

CHAPTER 3

Fractional Processes

A FRACTIONAL PROCESS SUBSTANTIALLY DIFFERS from a conventional one in its statistical properties. For instance, it may have a heavy-tailed probability distribution function (PDF), a hyperbolically decayed autocorrelation function (ACF), and a power spectrum density (PSD) function of $1/f$ type. It may have statistical dependence, either long-range dependence (LRD) or short-range dependence (SRD), and global or local self-similarity. This chapter gives the basics of fractional processes towards fractional random vibrations studied in this book.

3.1 INTRODUCTION

While considering random signals passing through vibration systems, we are interested in responses of fractional random vibrations driven by fractional processes. Fractional processes substantially differ from conventional ones in the following aspects.

1) Probability density function (PDF), $p(x)$, of a fractional process, $x(t)$, may be heavy tailed such that its mean or variance does not exist.

2) Autocorrelation function (ACF), $r_{xx}(\tau)$, of a fractional process $x(t)$, may be hyperbolically decayed such that it may be non-integrable for $t \in (0, \infty)$.

3) Power spectrum density (PSD) function, $S_{xx}(\omega)$, of a fractional process $x(t)$, may be divergent at $\omega = 0$.

DOI: 10.1201/9781003657897-3

In this chapter, we first discuss two numeric characteristics, namely, the Hurst parameter and fractal dimension (Mandelbrot [1], Li [2]). Then, a number of fractional processes are explained.

3.2 FRACTAL DIMENSION AND HURST PARAMETER

3.2.1 Fractal Dimension

In the field of fractional processes, we no longer pay attention to the numeric characteristics of mean and variance to characterize local and global properties of a random function as we do in the field of conventional random processes (Mandelbrot [1], Li [2]). Instead, we are interested in other two numeric characteristics, say, fractal dimension and the Hurst parameter, to describe local and global properties of a fractional process $x(t)$, respectively (Mandelbrot [1], Li [2], Li and Lim [3]). Let $r_{xx}(\tau)$ be the ACF of a fractional process $x(t)$. According to Kent and Wood [4], Hall and Roy [5], and Adler [6], if $r_{xx}(\tau)$ is sufficiently smooth on $(0, \infty)$ and if

$$r_{xx}(0) - r_{xx}(\tau) \sim c_1 |\tau|^{\alpha} \text{ for } |\tau| \to 0, \qquad (3.1)$$

where c_1 is a constant and $\alpha \in (0, 2)$ is the fractal index of $x(t)$, the fractal dimension of $x(t)$ is expressed by

$$D = 2 - \frac{\alpha}{2}. \qquad (3.2)$$

Fractal dimension D is a numeric characteristic to characterize a local property of $x(t)$ because it is attained when $\tau \to 0$. That local property is usually called local self-similarity or local roughness of $x(t)$. The larger the value of D the locally rougher of $x(t)$. When $\alpha \to 2$, $D \to 1$. If $\alpha \to 0$, $D \to 2$, implying that the fractal dimension of $x(t)$ approaches 2.

3.2.2 Hurst Parameter

If $x(t)$ is of long-range dependence (LRD), its ACF $r_{xx}(\tau)$ is non-integrable, that is,

$$\int_0^{\infty} r_{xx}(\tau) d\tau = \infty. \qquad (3.3)$$

On the other hand, if $\int_0^{\infty} r_{xx}(\tau) d\tau < \infty$, $x(t)$ is of short-range dependence (SRD).

A typical function form of such an ACF is a hyperbolically decayed function asymptotically expressed by

$$r_{xx}(\tau) \sim c|\tau|^{-\beta} \ (\tau \to \infty), \tag{3.4}$$

where $c > 0$ is a constant and $0 < \beta < 2$ is the dependence index. The previous expression implies a power law in the ACF of a fractional process. When $0 < \beta < 1$, β is an LRD index. If $1 < \beta < 2$, it is an SRD index.

In the field, β is usually expressed by the Hurst parameter $0 < H < 1$ to the memory of H. E. Hurst [7]. Thus,

$$\beta = 2 - 2H. \tag{3.5}$$

Therefore,

$$r_{xx}(\tau) \sim c|\tau|^{2H-2} \ (\tau \to \infty). \tag{3.6}$$

If $0.5 < H < 1$, we have $0 < \beta < 1$, implying $x(t)$ is of LRD. On the other side, $0 < H < 0.5 \leftrightarrow 1 < \beta < 2$, corresponding $x(t)$ is of SRD. Either LRD or SRD is a global property of $x(t)$. Thus, H is a numeric characteristic to measure a global property of $x(t)$ because it is obtained for $\tau \to \infty$.

Different from conventional random processes, we respectively use D and H to numerically characterize the local property and the global one of a fractional process $x(t)$ rather than mean and variance (Gneiting and Schlather [8], Lim and Li [9]).

We note that the estimation of H and or D becomes a branch of fractional processes as can be seen from Mandelbrot [10], Beran [11], Bassingthwaighte et al. [12], Taqqu et al. [13], Raymond et al. [14], Beran [15, 16], Cannon et al. [17], Caccia et al. [18, 19], Raymond and Bassingthwaighte [20, 21], Guerrero and Smith [22]), Veitch and Abry [23], Wornell [24], Abry et al. [25], Chen et al. [26], simply mentioning a few.

3.3 FRACTIONAL BROWNIAN MOTION AND FRACTIONAL GAUSSIAN NOISE

Fractional Brownian motion (fBm) and fractional Gaussian noise (fGn) are widely used fractional processes. The terms fBm and fGn are collective nouns. In this section, we shall discuss two kinds of fBms and fGns introduced by Mandelbrot and van Ness [27] (also Mandelbrot [28]), the fBm introduced by Kolmogorov, the fBm and fGn reported by Monin and Yaglom, and the generalized fGn (gfGn) established by Li.

3.3.1 FBms and fGns of Mandelbrot

Let $B(t)$ be the Brownian motion (Bm) (Hida [29]). Mandelbrot performed two kinds of fractional integrals, namely, the fractional integral of Weyl type and the fractional integral of Riemann-Liouville one, on $B(t)$ to introduce two kinds of fBms. Denote $B_{H,\,\text{Weyl}}(t)$ as the fBm when performing the fractional integral of Weyl type on $B(t)$. Let $B_{H,\,\text{RL}}(t)$ be the fBm when performing the fractional Riemann-Liouville integral on $B(t)$, referring to Miller and Ross [30] for the fractional integrals and derivatives of Weyl type and Riemann-Liouville one.

3.3.1.1 FBm of Weyl Type

Following Mandelbrot and van Ness [27], $B_{H,\,\text{Weyl}}(t)$ is given by

$$
B_{H,\text{Weyl}}(t) - B_{H,\text{Weyl}}(0) = \frac{1}{\Gamma(H+1/2)}\left\{ \int_{-\infty}^{0}\left[(t-u)^{H-0.5} - (-u)^{H-0.5}\right] \right.
$$
$$
\left. dB(u) + \int_{0}^{t}(t-u)^{H-0.5}\,dB(u)\right\},
$$

(3.7)

where $0 < H < 1$. Its ACF is expressed by

$$
r_{B_H,\text{Weyl}}(t,s) = \frac{V_H}{(H+1/2)\Gamma(H+1/2)}\left[|t|^{2H} + |s|^{2H} - |t-s|^{2H}\right], \quad (3.8)
$$

where V_H is the strength of $B_{H,\,\text{Weyl}}(t)$. It is given by

$$
V_H = \Gamma(1-2H)\frac{\cos \pi H}{\pi H}.
$$

(3.9)

Mandelbrot and van Ness did not mention the PSD of $B_{H,\,\text{Weyl}}(t)$ in their research [27]. However, Flandrin [31] introduced the PSD of $B_{H,\,\text{Weyl}}(t)$ in the form

$$
S_{B_H,\text{Weyl}}(t,\omega) = \frac{V_H}{|\omega|^{2H+1}}\left(1 - 2^{1-2H}\cos 2\omega t\right).
$$

(3.10)

The basic properties of $B_{H,\,\text{Weyl}}(t)$ are listed next.

- The fBm $B_{H,\,\text{Weyl}}(t)$ is non-stationary.

- The fBm $B_{H, Weyl}(t)$ is self-similar with the self-similarity measure $0 < H < 1$ as

$$B_{H, Weyl}(at) \equiv a^H B_{H, Weyl}(t), \quad a > 0, \tag{3.11}$$

where $\equiv$ denotes equality in the sense of probability distribution.

- When $0 < H < 1$ and $H \neq 0.5$, the PSD of $B_{H, Weyl}(t)$ is divergent at $\omega = 0$, exhibiting that it is a non-stationary $1/f$ noise.

- The property discussed earlier can be equivalently expressed like this. When $0 < H < 1$ and $H \neq 0.5$, $B_{H, Weyl}(t)$ is of LRD.

- The fBm $B_{H, Weyl}(t)$ is globally self-similar.

- The fBm $B_{H, Weyl}(t)$ reduces to the Bm if $H = 0.5$.

Let $D_{fBm, Weyl}$ be the fractal dimension of $B_{H, Weyl}(t)$. Then (Mandelbrot [1, 10]),

$$D_{fBm, Weyl} = 2 - H. \tag{3.12}$$

3.3.1.2 FBm of Riemann-Liouville Type

According to Mandelbrot and van Ness [27], $B_{H, RL}(t)$ is expressed by

$$B_{H, RL}(t) = {}_0 D_t^{-(H+1/2)} B'(t) = \frac{1}{\Gamma(H+1/2)} \int_0^t (t-u)^{H-1/2} dB(u). \tag{3.13}$$

The ACF of $B_{H, RL}(t)$ is given by

$$r_{B_H, RL}(t, s) = \frac{t^{H+1/2} s^{H-1/2}}{(H+1/2)\Gamma(H+1/2)^2} {}_2F_1(1/2 - H, 1, H+1/2, t/s), \tag{3.14}$$

where ${}_2F_1$ is the hypergeometric function (Lim and Muniandy [32]). The PSD of $B_{H, RL}(t)$ is in the form

$$S_{B_H, RL}(t, \omega) = \frac{\pi \omega t}{\omega^{2H+1}} [J_H(2\omega t) H_{H-1}(2\omega t) - J_{H-1}(2\omega t) H_H(2\omega t)], \tag{3.15}$$

where J_H is the Bessel function of order H, and H_H is the Struve function of order H (Lim and Muniandy [32]).

The properties of $B_{\mathrm{H,RL}}(t)$ are as follows.

- The fBm $B_{\mathrm{H,RL}}(t)$ is non-stationary.
- The fBm $B_{\mathrm{H,RL}}(t)$ is self-similar with the self-similarity measure $0 < H < 1$ since

$$B_{H,\mathrm{RL}}(at) \equiv a^H B_{H,\mathrm{RL}}(t), \quad a > 0, \tag{3.16}$$

where $\equiv$ denotes equality in probability distribution.

- When $0 < H < 1$ and $H \neq 0.5$, the PSD of $B_{\mathrm{H,RL}}(t)$ is divergent at $\omega = 0$, meaning that $B_{\mathrm{H,RL}}(t)$ is non-stationary $1/f$ noise.
- The fBm $B_{\mathrm{H,RL}}(t)$ is of LRD for $0 < H < 1$ and $H \neq 0.5$.
- The fBm $B_{\mathrm{H,RL}}(t)$ is globally self-similar.
- The fBm $B_{\mathrm{H,RL}}(t)$ reduces to the Bm when $H = 0.5$.

Let $D_{\mathrm{fBm,RL}}$ be the fractal dimension of $B_{\mathrm{H,Weyl}}(t)$. Then (Mandelbrot [1, 10]),

$$D_{\mathrm{fBm,RL}} = 2 - H. \tag{3.17}$$

3.3.1.3 FGn of Weyl Type

Let $G_{\mathrm{Weyl}}(t)$ be the fGn of Weyl type. It is the increment process of $B_{\mathrm{H,Weyl}}(t)$. Denote by $r_{\mathrm{fGn,Weyl}}(\tau; \varepsilon)$ the ACF of $G_{\mathrm{Weyl}}(t)$. Then (Mandelbrot and van Ness [27]),

$$r_{\mathrm{fGn,Weyl}}(\tau;\varepsilon) = \frac{V_H \varepsilon^{2H-2}}{2}\left[\left(\frac{|\tau|}{\varepsilon}+1\right)^{2H} + \left|\frac{|\tau|}{\varepsilon}-1\right|^{2H} - 2\left|\frac{\tau}{\varepsilon}\right|^{2H}\right], \tag{3.18}$$

where $0 < H < 1$ and $\varepsilon > 0$ is used by smoothing $B_{\mathrm{H,Weyl}}(t)$ so that the smoothed $B_{\mathrm{H,Weyl}}(t)$ is differentiable according to the theory of generalized functions.

Note that Mandelbrot and van Ness did not report the PSD of $G_{\mathrm{Weyl}}(t)$ in their research [27], but we obtained it (Li and Lim [33]). Let $S_{\mathrm{fGn,Weyl}}(\omega)$ be the PSD of $G_{\mathrm{Weyl}}(t)$. Then,

$$S_{\mathrm{fGn,Weyl}}(\omega) = V_H \sin(H\pi)\Gamma(2H+1)|\omega|^{1-2H}. \tag{3.19}$$

People usually use $\lim_{\varepsilon \to 0} r_{\mathrm{fGn,Weyl}}(\tau;\varepsilon)$ to represent the ACF of $G_{\mathrm{Weyl}}(t)$. Hence, the ACF of $G_{\mathrm{Weyl}}(t)$ is commonly written by

$$r_{\mathrm{fGn,Weyl}}(\tau) = \frac{V_H}{2}\left[\left(|\tau|+1\right)^{2H} + \left\||\tau|-1\right\|^{2H} - 2|\tau|^{2H}\right]. \qquad (3.20)$$

The function $r_{\mathrm{fGn,Weyl}}(\tau)$ is even. For $\tau > 0$, we have

$$r_{\mathrm{fGn,Weyl}}(\tau) = \frac{V_H}{2}\left[(\tau+1)^{2H} + (\tau-1)^{2H} - 2\tau^{2H}\right]. \qquad (3.21)$$

The right side of the previous equation is the finite second-order difference of $0.5(\tau)^{2H}$. Approximating it with the second-order differential of $0.5(\tau)^{2H}$ yields

$$r_{\mathrm{fGn,Weyl}}(\tau) \approx V_H H(2H-1)\tau^{2H-2}. \qquad (3.22)$$

The previous approximation is quite accurate for $\tau > 5$ (Li [2, 34, 35]). From the aforementioned, we can easily see that $G_{\mathrm{Weyl}}(t)$ includes three classes of processes. When $0.5 < H < 1$, $r_{\mathrm{fGn,Weyl}}(\tau)$ is non-integrable and accordingly $G_{\mathrm{Weyl}}(t)$ is of LRD. For $0 < H < 0.5$, the integral of $r_{\mathrm{fGn,Weyl}}(\tau)$ is convergent and consequently $G_{\mathrm{Weyl}}(t)$ is of SRD. The fGn $G_{\mathrm{Weyl}}(t)$ reduces to the white noise when $H = 0.5$.

Let $D_{\mathrm{fGn,Weyl}}$ be the fractal dimension of $G_{\mathrm{Weyl}}(t)$. Then (Mandelbrot [1, 10], Li [2, 34]),

$$D_{\mathrm{fGn,Weyl}} = 2-H. \qquad (3.23)$$

The quantity $D_{\mathrm{fGn,Weyl}}$ is correlated with H. Thus, $G_{\mathrm{Weyl}}(t)$ is self-similar with the self-similarity measure $0 < H < 1$ since

$$G_{\mathrm{Weyl}}(at) \equiv a^H B_{\mathrm{Weyl}}(t), \quad a > 0. \qquad (3.24)$$

Therefore, $G_{\mathrm{Weyl}}(t)$ is globally self-similar. Since H characterizes both the self-similarity and statistical dependences of $G_{\mathrm{Weyl}}(t)$, it is a single-parameter model. The fGn of Weyl type is only a stationary increment process with self-similarity (Samorodnitsky and Taqqu [36]).

3.3.1.4 FGn of Riemann-Liouville Type

Let $G_{RL}(t)$ be the fGn of Riemann-Liouville type. Then, it is the increment process of $B_{H,RL}(t)$. However, how to represent its ACF and or PSD remains challenging. For that reason, in applications, the term fGn is usually defaulted for $G_{Weyl}(t)$. Also, the term fBm is generally for $B_{H,Weyl}(t)$, see, for example, Bassingthwaighte et al. [12], Veitch and Abry [23], Wornell [24], Flandrin [31], Samorodnitsky and Taqqu [36], Kovin [37], Peters [38], and Levy-Vehel et al. [39].

3.3.2 FBm of Kolmogorov

When studying the Wiener spirals relating the Bm, Kolmogorov introduced an ACF, denoted by $B_\xi(\tau_1, \tau_2)$, in the form

$$B_\xi(\tau_1, \tau_2) = c\left(|\tau_1|^r + |\tau_2|^r - |\tau_1 - \tau_2|^r\right), \tag{3.25}$$

where $c \geq 0$ and $0 \leq r \leq 2$ (Kolmogorov [40, Theorem 6]). Although Kolmogorov did not mention the concept of fBm as people are discussing today by using fractional calculus, his work [40] happens to be consistent with the ACF of the fBm of Weyl type for $0 < r < 2$.

3.3.3 Structure-Function-Based fBm and fGn of Monin and Yaglom

Denote by $B_{MY}(t)$ the fBm stated by Monin and Yaglom in [41]. Without the generality losing, we assume that $B_{MY}(t)$ is mean zero. The variance of $[B_{MY}(t + \tau) - B_{MY}(t)]$ has the form (Monin and Yaglom [41, p. 83], Yaglom [42, p. 394])

$$\text{Var}[B_{MY}(t + \tau) - B_{MY}(t)] = S(\tau), \tag{3.26}$$

where Monin and Yaglom called $S(\tau)$ the structure function of $B_{MY}(t)$.

Let

$$S(\tau) = V_H|\tau|^{2H}. \tag{3.27}$$

Denote by $r_{MY}(\tau_1, \tau_2)$ the ACF of $B_{MY}(t)$. According to Li [35], Monin and Yaglom [41], Yaglom [42], Kaplan and Kuo [43],

$$r_{MY}(\tau_1, \tau_2) = \frac{1}{2}\left[S(\tau_1) + S(\tau_2) - S(|\tau_1 - \tau_2|)\right]. \tag{3.28}$$

Taking into account (3.27), we have

$$r_{MY}(\tau_1, \tau_2) = \frac{V_H}{2}\left[\left|\tau_1\right|^{2H} + \left|\tau_2\right|^{2H} - \left|\tau_1 - \tau_2\right|^{2H}\right].$$ (3.29)

Denote by $X_{MY}(t) = [B_{Yf}(t+1) - B_{MY}(t)]$ the increment function of $B_{MY}(t)$. Denote by $r_{XMY}(\tau)$ the ACF of $X_{MY}(t)$. Then,

$$r_{XMY}(\tau) = \frac{1}{2}\left[S(\tau+1) + S(\tau-1) - 2S(\tau)\right].$$ (3.30)

Considering (3.27), we have

$$r_{XMY}(\tau) = \frac{V_H}{2}\left[\left|\tau+1\right|^{2H} + \left|\tau-1\right|^{2H} - 2\left|\tau\right|^{2H}\right].$$ (3.31)

Note that the fBm and fGn of Monin and Yaglom are consistent with the fBm and fGn of Weyl type introduced by Mandelbrot in [27]. The difference between the two is in methodology. Mandelbrot utilized the tool of fractional calculus, but Monin and Yaglom did not.

3.3.4 Generalized fGn of Li

Recently, Li introduced a random function called the generalized fractional Gaussian noise (gfGn) in [44] and [45]. Let $0 < a \le 1$ be a constant. Let $C_{gfGn}(\tau)$ be the ACF of gfGn. Then (Li [44, Theorem 1]),

$$C_{gfGn}(\tau) = \frac{V_H}{2}\left[\left(\left|\tau^a\right|+1\right)^{2H} + \left|\left|\tau^a\right|-1\right|^{2H} - 2\left|\tau^a\right|^{2H}\right].$$ (3.32)

According to Li [44, Corollary 1], we have the approximation of $C_{gfGn}(\tau)$ in the form

$$C_{gfGn}(\tau) \approx V_H H(2H-1)\left|\tau\right|^{2aH-2a}.$$ (3.33)

Let $S_{gfGn}(\omega)$ be PSD of gfGn. Then (Li [44, Theorem 2], [46]),

$$S_{\text{gfGn}}(\omega) =$$

$$V_H \begin{cases} \begin{aligned} & \sin\left(Ha\pi\right)\Gamma(2Ha+1)\,|\omega|^{-2Ha-1} \\ & +0.5\sum_{k=0}^{\infty}\frac{[(-1)^{k+1}-1]\Gamma(2H+k)\Gamma(ak+1)}{\Gamma(2H)\Gamma(1+k)}\sin\left(\frac{ak\pi}{2}\right)|\omega|^{-ak-1} \qquad |\tau|<1, \\[10pt] & \sin\left(Ha\pi\right)\Gamma(2Ha+1)\,|\omega|^{-2Ha-1} \\ & +0.5\sum_{k=0}^{\infty}\frac{[(-1)^{k}-1]\Gamma(2H+k)\Gamma[(a(2H-k)+1]}{\Gamma(2H)\Gamma(1+k)}\sin\left[\frac{a(2H-k)\pi}{2}\right] \\ & |\omega|^{-a(2H-k)-1}, \quad |\tau|^{a}>1. \end{aligned} \end{cases} \qquad (3.34)$$

The approximation of $S_{\text{gfGn}}(\omega)$, following Li [44, Corollary 2], is given by

$$S_{\text{gfGn}}(\omega) \approx -V_H H(2H-1)\sin\left[a\pi(H-1)\right]\Gamma(2aH-2a+1)|\omega|^{-2a(H-1)-1}. \quad (3.35)$$

Let D_{gfGn} be the fractal dimension of gfGn. Then,

$$D_{\text{gfGn}} = 2 - aH. \qquad (3.36)$$

The fractal dimension D_{gfGn} is correlated with H with the coupling factor a.

Note that the gfGn of Li reduces to the fGn of Weyl type if $a = 1$. The model of gfGn has been applied recently, see, for example, Avraham and Pinchas [47–49] and Sousa-Vieira and Fernández-Veiga [50].

3.4 GENERALIZED CAUCHY PROCESS

Following Li [2], Li and Lim [3], Gneiting and Schlather [8], Lim and Li [9], and Li [34, 35, 51–53], we say that $x(t)$ is a generalized Cauchy (GC) process when its ACF follows.

$$C_{\text{GC}}(\tau) = \mathrm{E}\left[x(t+\tau)x(t)\right] = \psi^2\left(1+|\tau|^{\alpha}\right)^{-\frac{\beta}{\alpha}}, \qquad (3.37)$$

where $0 < \alpha \le 2$, $\beta > 0$, and ψ^2 is the strength of $x(t)$. The ACF $C_{\text{GC}}(\tau)$ is positive-definite for the above ranges of α and β. It is a completely monotone for $0 < \alpha \le 1$ and $\beta > 0$. Without losing the generality, let $\psi^2 = 1$ in what follows. When $\alpha = \beta = 2$, one gets the usual Cauchy process that is modelled by its ACF in the form

$$C(\tau) = \left(1 + |\tau|^2\right)^{-1}, \tag{3.38}$$

which has been applied to geostatistics, see, for example, Chiles and Delfiner [54].

Considering a GC process globally and locally, we have the following asymptotic expressions

$$C_{GC}(0) - C_{GC}(\tau) \sim |\tau|^{\alpha}, \quad \tau \to 0, \tag{3.39}$$

$$C_{GC}(\tau) \sim |\tau|^{-\beta}, \quad \tau \to \infty. \tag{3.40}$$

Therefore, the fractal dimension and the Hurst parameter of a GC process are respectively expressed by

$$D_{GC} = 2 - \frac{\alpha}{2}, \tag{3.41}$$

$$H_{GC} = 1 - \frac{\beta}{2}. \tag{3.42}$$

Since $0 < \alpha \leq 2$ (Hall and Roy [5]), we have

$$1 \leq D_{GC} < 2. \tag{3.43}$$

As $0 < H_{GC} < 1$ (Beran [11]), we have

$$0 < \beta < 2. \tag{3.44}$$

Because $0 < \beta < 1$ corresponds to $0.5 < H_{GC} < 1$, a GC process is of LRD for $0 < \beta < 1$. Since $1 < \beta < 2$ corresponds to $0 < H_{GC} < 0.5$, a GC process is of SRD for $1 < \beta < 2$.

Let $Sa(\omega) = (\sin\omega)/\omega$. Then, the PSD of the GC process is given by (Li and Lim [55])

$$S_{GC}(\omega) = \sum_{k=0}^{\infty} \frac{(-1)^k \Gamma[(\beta/\alpha) + k]}{\pi \Gamma(\beta/\alpha) \Gamma(1+k)} I_1(\omega)^* Sa(\omega)$$
$$+ \sum_{k=0}^{\infty} \frac{(-1)^k \Gamma[(\beta/\alpha) + k]}{\pi \Gamma(\beta/\alpha) \Gamma(1+k)} \left[\pi I_2(\omega) - I_2(\omega)^* Sa(\omega) \right], \tag{3.45}$$

where

$$I_1(\omega) = -2\sin(\alpha k\pi/2)\Gamma(\alpha k+1)|\omega|^{-\alpha k-1}, \tag{3.46}$$

$$I_2(\omega) = 2\sin[(\beta+\alpha k)\pi/2]\Gamma[1-(\beta+\alpha k)]|\omega|^{(\beta+\alpha k)-1}. \tag{3.47}$$

With (3.41) and (3.42) for the expressions of D_{GC} and H_{GC}, we may write $C_{\text{GC}}(\tau)$ by

$$C_{\text{GC}}(\tau) = \left(1+|\tau|^{4-2D_{\text{GC}}}\right)^{-\frac{1-H_{\text{GC}}}{2-D_{\text{GC}}}}, \tag{3.48}$$

and $S_{\text{GC}}(\omega)$ by

$$S_{\text{GC}}(\omega) = \sum_{k=0}^{\infty} A_1(D_{\text{GC}}, H_{\text{GC}}, k)|\omega|^{-(4-2D_{\text{GC}})k-1} * \text{Sa}(\omega)$$

$$+\sum_{k=0}^{\infty} A_2(D_{\text{GC}}, H_{\text{GC}}, k)\left\{\pi|\omega|^{2[0.5-H_{\text{GC}}+(2-D_{\text{GC}})k]} - |\omega|^{2[0.5-H_{\text{GC}}+(2-D_{\text{GC}})k]} * \text{Sa}(\omega)\right\},$$

$$\tag{3.49}$$

where

$$A_1(D_{\text{GC}}, H_{\text{GC}}, k) =$$

$$\frac{(-1)^{k+1}2\Gamma\left(\frac{1-H_{\text{GC}}}{2-D_{\text{GC}}}+k\right)\sin[(2-D_{\text{GC}})k\pi]\Gamma[(4-2D_{\text{GC}})k+1]}{\pi\Gamma\left(\frac{1-H_{\text{GC}}}{2-D_{\text{GC}}}\right)\Gamma(1+k)}, \tag{3.50}$$

and

$$A_2(D_{\text{GC}}, H_{\text{GC}}, k)$$

$$= \frac{(-1)^k 2\Gamma\left(\frac{1-H_{\text{GC}}}{2-D_{\text{GC}}}+k\right)\sin\{[1-H_{\text{GC}}+(2-D_{\text{GC}})k]\pi\}\Gamma\{2[H_{\text{GC}}-0.5-(2-D_{\text{GC}})k]\}}{\pi\Gamma\left(\frac{1-H_{\text{GC}}}{2-D_{\text{GC}}}\right)\Gamma(1+k)}. \tag{3.51}$$

Because the GC process is of LRD for $0 < \beta < 1$ and SRD for $1 < \beta < 2$, its statistical dependences are only measured by H. The GC process is not globally self-similar but local self-similar with the similarity measure D_{GC}. In short, the GC model can be used to decouple the local behaviour and the global one of fractional processes, being flexibly better agreement with the data for both short-term and long-term lags when the local self-similarity and the LRD need studying separately, such as network traffic (Li [2, 35]). Some applications of the GC process refer to Sousa-Vieira and Fernández-Veiga [50], Li and Li [53], Lim and Teo [56], Vengadesh et al. [57], Muniandy and Stanslas [58, 59], and Asgari et al. [60].

3.5 FRACTIONAL ORNSTEIN-UHLENBECK PROCESS

3.5.1 Ordinary Ornstein-Uhlenbeck (OU) Process

The following

$$\begin{cases} \left(\dfrac{d}{dt} + \lambda\right) x(t) = w(t), \\ x(0) = x_0, \end{cases} \tag{3.52}$$

is termed the Langevin equation, where λ is a positive parameter, $w(t)$ is the white noise with mean zero, and x_0 is a random variable independent of the standard Bm $B(t)$. The solution of the Langevin equation is called the OU process, according to the work by Uhlenbeck and Ornstein on Brownian motion [61]. The solution of the Langevin equation is given by

$$x(t) = x_0 e^{-\lambda t} + \int_{-\infty}^{t} e^{\lambda u} w(u) du. \tag{3.53}$$

Let the Fourier transforms of $w(t)$ and $x(t)$ be $W(\omega)$ and $X(\omega)$, respectively. Denote by the frequency transfer function of the Langevin equation by $H_{OU}(\omega)$. Then,

$$H_{OU}(\omega) = \frac{1}{\lambda + i\omega}. \tag{3.54}$$

Therefore,

$$X(\omega) = H_{OU}(\omega) W(\omega). \tag{3.55}$$

For normalized $w(t)$, $|W(\omega)|^2 = 1$. Denote by $S_{OU}(\omega)$ the PSD of OU process. Then,

$$S_{OU}(\omega) = \frac{1}{\lambda^2 + \omega^2}. \tag{3.56}$$

Denote by $r_{OU}(\tau)$ the ACF of OU process. Then,

$$r_{OU}(\tau) = \mathbf{F}^{-1}[S_{OU}(\omega)] = \frac{e^{-\lambda|\tau|}}{2\lambda}, \tag{3.57}$$

where $\mathbf{F}^{-1}$ is the operator of the inverse Fourier transform.

OU process is of SRD since $S_{OU}(0)$ is convergent.

3.5.2 Fractional OU Process

Let $\beta > 0$. Consider the following fractional Langevin equation:

$$\left(\frac{d}{dt} + \lambda\right)^{\beta} x_1(t) = w(t). \tag{3.58}$$

Denote by $h_{x_1}(t)$ the impulse response function of the previous system. It is the solution to the following equation:

$$\left(\frac{d}{dt} + \lambda\right)^{\beta} h_{x_1}(t) = \delta(t), \tag{3.59}$$

where $\delta(t)$ is the delta function. Let $H_{x_1}(\omega)$ be the Fourier transform of $h_{x_1}(t)$. Then,

$$H_{x_1}(\omega) = \frac{1}{(\lambda - i\omega)^{\beta}}. \tag{3.60}$$

Let $S_{x_1}(\omega)$ be the PSD of the fractional OU process $x_1(t)$. For normalized $w(t)$, we have

$$S_{x_1}(\omega) = \frac{1}{(\lambda^2 + \omega^2)^{\beta}}. \tag{3.61}$$

Let $r_{x_1}(\tau)$ be the ACF of $x_1(t)$. Then,

$$r_{x_1}(\tau) = F^{-1}\left[S_{x_1}(\omega)\right] = \frac{\lambda^{-2\nu}}{2^\nu \sqrt{\pi}\Gamma(\nu+1/2)}|\lambda\tau|^\nu K_\nu\left(|\lambda\tau|\right), \qquad (3.62)$$

where $\nu = \beta - 1/2$ and K_ν is the modified Bessel function of the second kind of order ν (Gelfand and Vilenkin [62], Lim and Muniandy [63], Lim et al [64, 65]).

Let $\nu = H \in (0, 1)$. Then,

$$S_{x_1}(\omega) = \frac{1}{\left(\lambda^2 + \omega^2\right)^{H+1/2}}. \qquad (3.63)$$

From this, we see that $x_1(t)$ is of SRD because its PSD is convergent for $\omega \to 0$.

If $\omega \gg \lambda$, we have

$$S_{x_1}(\omega) \sim \frac{1}{|\omega|^{2H+1}}. \qquad (3.64)$$

Therefore, for large ω, the ACF of the fractional OU process $x_1(t)$ may be approximately expressed by

$$r_{x_1}(\tau) \approx F^{-1}\left(\frac{1}{|\omega|^{2H+1}}\right) \sim c|\tau|^{2H}, \qquad (3.65)$$

where c is a constant. The fractal dimension of $x_1(t)$ is given by

$$D_{X_1} = 2 - H. \qquad (3.66)$$

3.5.3 Von Kármán Spectrum Viewed from Fractional OU Process

Consider the following fractional Langevin equation

$$\sqrt{A_{vk}}\left(\frac{d}{dt} + B_{vk}\right)^{5/6} X_{vk}(t) = w(t), \qquad (3.67)$$

where

$$A_{vk} = \frac{4\sigma_u^2}{70.8^{5/6}\left(\dfrac{L_u^x}{U}\right)^{2/3}} \tag{3.68}$$

and

$$B_{vk} = \frac{U}{70.8^{1/2}\,L_u^x}, \tag{3.69}$$

where L_u^x is turbulence integral scale, U is mean speed, u_f is friction velocity, b_v is friction velocity coefficient such that the variance of wind speed $\sigma_u^2 = b_v u_f^2$.

Denote by $S_{vk}(\omega)$ the PSD of $X_{vk}(t)$. Then,

$$S_{vk}(\omega) = \frac{A_{vk}}{\left[(B_{vk})^2 + \omega^2\right]^{5/6}}. \tag{3.70}$$

Let $r_{vk}(\tau)$ be the ACF of $X_{vk}(t)$. Then,

$$r_{vk}(\tau) = \frac{2\sqrt{\pi}\,A_{vk}}{\Gamma(5/6)(B_{vk})^{1/3}}\left(\frac{|\tau|}{2}\right)^{1/3} K_{1/3}\left(B_{vk}\,|\tau|\right), \tag{3.71}$$

where $K_{1/3}(\cdot)$ is the modified Bessel function of second kind $K_v(z)$ for $v = 1/3$.

Because

$$r_{vk}(\tau) \sim |\tau|^{2/3} \quad \text{for } \tau \to 0, \tag{3.72}$$

we have

$$r_{vk}(0) - r_{vk}(\tau) \sim \left(|\tau|^a\right)\Big|_{a=2/3} \quad \text{for } \tau \to 0. \tag{3.73}$$

With the probability one, the fractal dimension of the von Kármán process is given by

$$D_{vk} = \left(2 - \frac{a}{2}\right)\Bigg|_{a=2/3} = \frac{5}{3}. \tag{3.74}$$

The von Kármán process is of SRD, but it is a fractional one (Li [66]).

3.6 SUMMARY

Several models, fBms, fGns, the GC process, fractional OU process, and the von Kármán process have been addressed. Fractal dimension and the Hurst parameter have been paid attention to. Local self-similarity and statistical dependences (LRD and SRD) of fractional processes have been emphasized. The von Kármán process used in wind engineering has been explained from a view of fractional processes. The concepts, such as $1/f$ noise, power laws in PDF, ACF, and PSD in fractional processes, have been mentioned.

3.7 EXERCISES

3.1. Let the ACF of a random function $x(t)$ be $r(\tau) = \left(1 + |\tau|^2\right)^{-1}$. Prove that $x(t)$ is of SRD.

3.2. Let the PSD of a random function $x(t)$ be $S(\omega) = \dfrac{1}{\sqrt{2\pi}} |\omega|^{1/2} K_{1/2}\left(|\omega|\right)$, where $K_v(\cdot)$ is the modified Bessel function of the second kind of order v. Prove that $x(t)$ is of SRD.

3.3. Let the PSD of a random function $x(t)$ be $S(\omega) = \dfrac{1}{|\omega|^{2H+1}}$ $\left(1 - 2^{1-2H} \cos 2\omega t\right)$. Prove that $x(t)$ is of LRD for $0 < H < 0.5$ and $0.5 < H < 1$.

3.4. Let the PSD of a random function $x(t)$ be $S(\omega) = \sin(H\pi)\Gamma(2H+1)$ $|\omega|^{1-2H}$. Prove that $x(t)$ is of SRD for $0 < H < 0.5$ and LRD if $0.5 < H < 1$.

3.5. Find the inverse Fourier transform of $S(\omega) = \sin(H\pi)\Gamma(2H+1)|\omega|^{1-2H}$, where $0 < H < 1$.

3.6. Prove that the GC process is non-Markovian.

3.7. Prove that the fGn of Riemann-Liouville type is non-stationary.

3.8. Let the PDF of a random function $x(t)$ be $p(x) = \dfrac{ab}{x^{a+1}}$, where $x \geq a$. Prove that the mean and variance of $x(t)$ do not exist when $a = 1$.

3.9. Let the PDF of a random function $x(t)$ be $p(x) = \dfrac{1}{\pi\left(1+x^2\right)}$. Prove that the variance of $x(t)$ does not exist.

3.10. Let the PSD of a random function $x(t)$ be $p(x) = \dfrac{1}{\pi\left(1+x^2\right)}$. Prove that $x(t)$ is with zero mean.

3.11. Let $S(\omega)$ be the PSD of $x(t)$. Prove that $x(t)$ is of LRD if $S(0) = \infty$.

3.12. Let $S(\omega)$ be the PSD of $x(t)$. Prove that $x(t)$ is of SRD if $S(0) < \infty$.

3.13. Prove that fBm is of LRD for $0 < H < 0.5$ and $0.5 < H < 1$.

REFERENCES

1. B. B. Mandelbrot, *The Fractal Geometry of Nature*, W. H. Freeman, New York, 1982.
2. M. Li, *Fractal Teletraffic Modeling and Delay Bounds in Computer Communications*, CRC Press, Boca Raton, 2022.
3. M. Li and S. C. Lim, Modeling network traffic using generalized Cauchy process, *Physica A*, vol. 387, no. 11, 2008, 2584–2594.
4. J. T. Kent and A. T. Wood, Estimating the fractal dimension of a locally self-similar Gaussian process by using increments, *Journal of the Royal Statistical Society B*, vol. 59, no. 3, 1997, 679–699.
5. P. Hall and R. Roy, On the Relationship between fractal dimension and fractal index for stationary stochastic processes, *The Annals of Applied Probability*, vol. 4, no. 1, 1994, 241–253.
6. A. J. Adler, *The Geometry of Random Fields*, John Wiley & Sons, New York, 1981.
7. H. E. Hurst, Long term storage capacity of reservoirs, *Transactions of the American Society of Civil Engineers*, vol. 116, 1951, 770–799.
8. T. Gneiting and M. Schlather, Stochastic models that separate fractal dimension and Hurst effect, *SIAM Review*, vol. 46, no. 2, 2004, 269–282.
9. S. C. Lim and M. Li, Generalized Cauchy process and its application to relaxation phenomena, *Journal of Physics A: Mathematical and General*, vol. 39, no. 2, 2006, 2935–2951.
10. B. B. Mandelbrot, *Gaussian Self-Affinity and Fractals*, Springer, New York, 2001.
11. J. Beran, *Statistics for Long-Memory Processes*, Chapman & Hall, New York, 1994.
12. J. B. Bassingthwaighte, L. S. Liebovitch, and B. J. West, *Fractal Physiology*, Oxford University Press, Oxford, 1994.

13. M. S. Taqqu, V. Teverovsky, and W. Willinger, Estimators for long-range dependence: An empirical study, *Fractals*, vol. 3, no. 4, 1995, 785–798.
14. G. M. Raymond, D. B. Percival, J. B. Bassingthwaighte, The spectra and periodograms of anti-correlated discrete fractional Gaussian noise, *Physica A*, vol. 322, 2003, 169–179.
15. J. Beran, Fitting long-memory models by generalized linear regression, *Biometrika*, vol. 80, no. 4, 1993, 817–822.
16. J. Beran, On parameter estimation for locally stationary long-memory processes, *Journal of Statistical Planning and Inference*, vol. 139, no. 3, 2009, 900–915.
17. M. J. Cannon, D. B. Percival, D. C. Caccia, G. M. Raymond, and J. B. Bassingthwaighte, Evaluating scaled windowed variance methods for estimating the Hurst coefficient of time series, *Physica A*, vol. 241, no. 3–4, 1997, 606–626.
18. D. C. Caccia, D. B. Percival, M. J. Cannon, G. M. Raymond, and J. B. Bassingthwaighte, Analyzing exact fractal time series: Evaluating scaled windowed variance methods for estimating the Hurst coefficient of time series, *Physica A*, vol. 246, no. 3–4, 1997, 609–632.
19. D. C. Caccia, D. Percival, M. J. Cannon, G. Raymond, and J. B. Bassingthwaighte, Evaluating maximum likelihood estimation methods to determine the Hurst coefficient, *Physica A*, vol. 273, no. 3–4, 1999, 439–451.
20. J. B. Bassingthwaighte and G. M. Raymond, Evaluating rescaled range analysis for time series, *Annals of Biomedical Engineering*, vol. 22, no. 4, 1994, 432–444.
21. G. M. Raymond and J. B. Bassingthwaighte, Deriving dispersional and scaled windowed variance analyses of the correlation of discrete fractional Gaussian noise, *Physica A*, vol. 265, no. 1–2, 1999, 85–96.
22. A. Guerrero and L. A. Smith, A maximum likelihood estimator for long-range persistence, *Physica A*, vol. 355, no. 2–4, 2005, 619–632.
23. D. Veitch and P. Abry, Wavelet-based joint estimate of the long-range dependence parameters, *IEEE Transactions Information Theory*, 45, no. 3, 1999, 878–897.
24. G. W. Wornell, Wavelet-based representations for the 1 over f family of fractal processes, *Proceedings of the IEEE*, vol. 81, no. 10, 1993, 1428–1450.
25. P. Abry, D. Veitch, and P. Flandrin, Long-range dependence—revisiting aggregation with wavelets, *Journal of Time Series Analysis*, vol. 19, no. 3, 1998, 253–266.
26. Y.-Q. Chen, R. Sun, and A. Zhou, An improved Hurst parameter estimator based on fractional Fourier transform, *Telecommunication Systems*, vol. 43, no. 3–4, 2010, 197–206.
27. B. B. Mandelbrot and J. W. van Ness, Fractional Brownian motions, fractional noises and applications, *SIAM Review*, vol. 10, no. 4, 1968, 422–437.
28. B. B. Mandelbrot, Note on the definition and the stationarity of fractional Gaussian noise, *Journal of Hydrology*, vol. 30, no. 4, 1976, 407–409.
29. T. Hida, *Brownian Motion*, Springer, Berlin, 1980.

30. K. S. Miller and B. Ross, *An Introduction to the Fractional Calculus and Fractional Differential Equations*, John Wiley & Sons, New York, 1993.
31. P. Flandrin, On the spectrum of fractional Brownian motion, *IEEE Transactions of Information Theory*, vol. 35, no. 1, 1989, 197–199.
32. S. C. Lim and S. V. Muniandy, On some possible generalizations of fractional Brownian motion, *Physics Letters A*, vol. 226, no. 2–3, 2000, 140–145.
33. M. Li and S. C. Lim, A rigorous derivation of power spectrum of fractional Gaussian noise, *Fluctuation and Noise Letters*, vol. 6, no. 4, 2006, C33–C36.
34. M. Li, Modified multifractional Gaussian noise and its application, *Physica Scripta*, vol. 96, no. 12, 2021, 125002 (12 pages).
35. M. Li, *Multi-Fractal Traffic and Anomaly Detection in Computer Communications*, CRC Press, Boca Raton, 2022.
36. G. Samorodnitsky and M. S. Taqqu, *Stable Non-Gaussian Random Processes: Stochastic Models with Infinite Variance*, Chapman and Hall, New York, 1994.
37. G. Korvin, *Fractal Models in the Earth Science*, Elsevier, Amsterdam, The Netherlands, 1992.
38. E. E. Peters, *Fractal Market Analysis—Applying Chaos Theory to Investment and Economics*, John Wiley & Sons, New York, 1994.
39. J. Levy-Vehel, E. Lutton, and C. Tricot, editors, *Fractals in Engineering*, Springer, Berlin, 1997.
40. A. N. Kolmogorov, Wiener spirals and some other interesting curves in the Hilbert space, *Academy of Science URSS*, vol. 26, no. 2, 1940, 115–118. English translation.
41. A. S. Monin and A. M. Yaglom, *Statistical Fluid Mechanics: Mechanics of Turbulence*, vol. 2, English Ed., ed. by John L. Lumley, The MIT Press, Cambridge, MA, 1971.
42. A. M. Yaglom, *Correlation Theory of Stationary and Related Random Functions, Vol. I: Basic Results*, Springer, New York, 1987.
43. L. M. Kaplan and C. -C. J. Kuo, Extending self-similarity for fractional Brownian motion, *IEEE Transactions on Signal Processing*, vol. 42, no. 12, 1994, 3526–3530.
44. M. Li, Generalized fractional Gaussian noise and its application to traffic modeling, *Physica A*, vol. 579, 2021, 1236137 (22 pages).
45. M. Li, Modeling autocorrelation functions of long-range dependent teletraffic series based on optimal approximation in Hilbert space-a further study, *Applied Mathematical Modelling*, vol. 31, no. 3, 2007, 625–631.
46. M. Li, Power spectrum of generalized fractional Gaussian noise, *Advances in Mathematical Physics*, vol. 2013, Article ID 315979, 3 pages, 2013.
47. Y. Avraham and M. Pinchas, A novel clock skew estimator and its performance for the IEEE 1588v2 (PTP) Case in fractional Gaussian noise/generalized fractional Gaussian noise environment, *Frontiers in Physics*, vol. 9, 2021, 796811.
48. Y. Avraham and M. Pinchas, Two novel one-way delay clock skew estimators and their performances for the fractional Gaussian noise/generalized fractional Gaussian noise environment applicable for the IEEE 1588v2 (PTP) case, *Frontiers in Physics*, vol. 10, 2022, 867861.

49. Y. Avraham and M. Pinchas, A low-computational burden closed-form approximated expression for MSE applicable for PTP with gfGn environment, *Fractal and Fractional*, vol. 8, no. 7, 2024, 418.

50. M. E. Sousa-Vieira and M. Fernández-Veiga, Efficient generators of the generalized fractional Gaussian noise and Cauchy processes, *Fractal and Fractional*, vol. 7, no. 6, 2023, 455.

51. M. Li, Long-range dependence and self-similarity of teletraffic with different protocols at the large time scale of day in the duration of 12 years: Autocorrelation modeling, *Physica Scripta*, vol. 95, no. 4, 2020, 065222, (15 pages).

52. M. Li, Multi-fractional generalized Cauchy process and its application to teletraffic, *Physica A*, vol. 550, 2020, 123982 (14 pages).

53. M. Li and J.-Y. Li, Generalized Cauchy model of sea level fluctuations with long-range dependence, *Physica A*, vol. 484, 2017, 309–335.

54. J.-P. Chiles and P. Delfiner, *Geostatistics, Modeling Spatial Uncertainty*, John Wiley & Sons, New York, 1999.

55. M. Li and S. C. Lim, Power spectrum of generalized Cauchy process, *Telecommunication Systems*, vol. 43, no. 3–4, 2010, 219–222.

56. S. C. Lim and L. P. Teo, Gaussian fields and Gaussian sheets with generalized Cauchy covariance structure, *Stochastic Processes and Their Applications*, vol. 119, no. 4, 2009, 1325–1356.

57. P. Vengadesh, S. V. Muniandy, and W. H. Abd. Majid, Fractal morphological analysis of Bacteriorhodopsin (bR) layers deposited onto Indium Tin Oxide (ITO) electrodes, *Materials Science and Engineering: C*, vol. 29, no. 5, 2009, 1621–1626.

58. S. V. Muniandy and J. Stanslas, Modelling of chromatin morphologies in breast cancer cells undergoing apoptosis using generalized Cauchy field, *Computerized Medical Imaging and Graphics*, vol. 32, no. 7, 2008, 631–637.

59. S. V. Muniandy, W. X. Chew, and C. S. Wong, Fractional dynamics in the light scattering intensity fluctuation in dusty plasma, *Physics of Plasmas*, vol. 18, no. 1, 2011, 013701.

60. H. Asgari, S. V. Muniandy, and C. S. Wong, Stochastic dynamics of charge fluctuations in dusty plasma: A non-Markovian approach, *Physics of Plasmas*, vol. 18, no. 8, 2011, 083709.

61. G. E. Uhlenbeck and L. S. Ornstein, On the theory of the Brownian motion, *Physical Review*, vol. 36, no. 5, 1930, 823–841.

62. I. M. Gelfand and K. Vilenkin, *Generalized Functions*, vol. 1, Academic Press, New York, 1964.

63. S. C. Lim and S. V. Muniandy, Generalized Ornstein-Uhlenbeck processes and associated self-similar processes, *Journal of Physics A: Mathematical and General*, vol. 36, no. 14, 2003, 3961–3982.

64. S. C. Lim, M. Li, and L. P. Teo, Locally self-similar fractional oscillator processes, *Fluctuation and Noise Letters*, vol. 7, no. 2, 2007, L169–L179.

65. S. C. Lim, M. Li, and L. P. Teo, Langevin equation with two fractional orders, *Physics Letters A*, vol. 372, no. 42, 2008, 6309–6320.

66. M. Li, *Fractional Vibrations with Applications to Euler-Bernoulli Beams*, CRC Press, Boca Raton, 2023.

CHAPTER 4

Responses of Random Vibration Systems and Input-Output Relations

THIS CHAPTER DISCUSSES THE foundation of random signals passing through linear vibrators. The main contents are the input-output relationships of random vibration systems.

4.1 BACKGROUND

Consider a linear vibration system with the primary mass m, damping c, and stiffness k. Its vibration motion equation is in the form

$$mx''(t) + cx'(t) + kx(t) = p(t). \tag{4.1}$$

In (4.1), $p(t)$ is a driven signal and $x(t)$ is response. A vibration system can be characterized by its impulse response function $h(t)$ in the time domain or its frequency transfer function $H(\omega) \leftrightarrow h(t)$ in the frequency domain.

In the time domain, the input-output relationship between $p(t)$ and $x(t)$ is given by

$$x(t) = p(t) * h(t). \tag{4.2}$$

In (4.2), $*$ stands for the operation of convolution. In the frequency domain, the input-output relationship between $P(\omega) \leftrightarrow p(t)$ and $X(\omega) \leftrightarrow x(t)$ is expressed by (4.3) in the form

DOI: 10.1201/9781003657897-4

$$X(\omega) = H(\omega)P(\omega). \tag{4.3}$$

When a driven force $p(t)$ is random, response $x(t)$ is random too. As the Fourier transform of a random signal is random (Robinson [1]), $P(\omega)$ and $X(\omega)$ are random if $p(t)$ is random. When a driven signal is random, the vibration equation becomes a stochastic differential equation. For a stochastic vibration equation (4.1), taking random function $x(t)$ as a solution does not make sense. Hence, (4.2) and (4.3) are not enough for random vibrations. The meaningful solution is a stochastic model of $x(t)$, for example, probability density function (PDF), autocorrelation function (ACF), power spectrum density (PSD) function, or cross-correlation or cross PSD between $x(t)$ and $p(t)$.

In random vibrations, due to the importance of frequency, one is interested in the PSD of $x(t)$ or equivalently in mathematics, ACF of $x(t)$. In addition, one is interested in the cross-PSD response or the cross-correlation response. Thus, we need to establish the relationship between the PSD excitation $S_{pp}(\omega)$ (equivalently, the ACF excitation $r_{pp}(\tau)$) and PSD response $S_{xx}(\omega)$ (equivalently, the ACF response $r_{xx}(\tau)$). Additionally, we need the relationship between the PSD excitation $S_{pp}(\omega)$ (or the ACF excitation $r_{pp}(\tau)$) and the cross-PSD response $S_{px}(\omega)$ (or the cross-correlation response $r_{px}(\tau)$). Figure 4.1 is a diagram to abstractly describe a random vibration system. Note that zero-frequency response and mean square response are useful in practice.

The rest of the chapter is organized as follows. PSD response and PSD-based input-output relation are discussed in Section 4.2. ACF response and ACF-based input-output relation are explained in Section 4.3. Zero-frequency response is in Section 4.4. Mean square response is given in Section 4.5. Section 4.6 gives a case study. Section 4.7 depicts the cross-correlation response and cross-correlation-based input-output relation. The cross-PSD response and cross-PSD-based input-output relation are addressed in Section 4.8. Coherence function is described in Section 4.9, which is followed by summary.

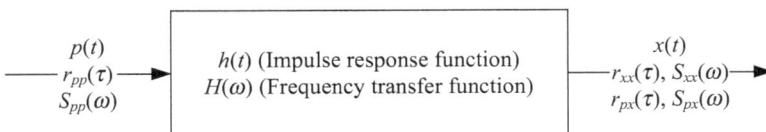

FIGURE 4.1   Diagram of a random vibration system.

4.2 PSD RESPONSE AND PSD-BASED INPUT-OUTPUT RELATION

From (4.3), one has the expression of $|X(\omega)|$ expressed by (4.4) in the form

$$|X(\omega)| = |H(\omega)P(\omega)|. \qquad (4.4)$$

From (4.4), we have the expression of $|X(\omega)|^2$ expressed by (4.5).

$$|X(\omega)|^2 = |H(\omega)|^2|P(\omega)|^2. \qquad (4.5)$$

Since $S_{pp}(\omega) = |P(\omega)|^2$ and $S_{xx}(\omega) = |X(\omega)|^2$, the following (4.6) results.

$$S_{xx}(\omega) = |H(\omega)|^2 S_{pp}(\omega). \qquad (4.6)$$

Eq. (4.6) is an input-output relation between $S_{xx}(\omega)$ and $S_{pp}(\omega)$.

4.3 ACF RESPONSE AND ACF-BASED INPUT-OUTPUT RELATION

Note that

$$|H(\omega)|^2 = H(\omega)H^*(\omega). \qquad (4.7)$$

In (4.7), $H^*(\omega)$ is the complex conjugate of $H(\omega)$. Eq. (4.8) implies that $H(\omega)H^*(\omega)$ and $h(t)^*h(-t)$ is a pair of Fourier transform,

$$H(\omega)H^*(\omega) \leftrightarrow h(t)^* h(-t). \qquad (4.8)$$

Thus, doing the inverse Fourier transform on both sides of (4.6) produces

$$r_{xx}(\tau) = h(\tau)^* h(-\tau)^* r_{pp}(\tau). \qquad (4.9)$$

Eq. (4.9) is the ACF-based input-output relation.

4.4 ZERO-FREQUENCY RESPONSE

We call $S_{xx}(0)$ zero-frequency response. It is expressed by (4.10).

$$S_{xx}(0) = |H(0)|^2 S_{pp}(0). \qquad (4.10)$$

Equivalently, one may use (4.11) to express $S_{xx}(0)$ by

$$S_{xx}(0) = \int_{-\infty}^{\infty} r_{xx}(\tau)d\tau. \tag{4.11}$$

4.5 MEAN SQUARE RESPONSE

Denote by Ψ_{xx}^2 the mean square of the response $x(t)$. It is also the mean square response. Eq. (4.12) is its expression.

$$\Psi_{xx}^2 = r_{xx}(0) = \int_{-\infty}^{\infty} S_{xx}(\omega)d\omega = \int_{-\infty}^{\infty} |H(\omega)|^2 S_{pp}(\omega)d\omega. \tag{4.12}$$

Since $r_{xx}(\tau) = E[x(t)x(t+\tau)]$, we have the mean square response in the form of (4.13).

$$r_{xx}(0) = E[x^2(t)]. \tag{4.13}$$

4.6 CASE STUDY

Consider an ocean wave signal passing through a vibration system with the impulse response

$$h(t) = \frac{1}{m\omega_d} e^{-\varsigma\omega_n t} \sin \omega_d t u(t). \tag{4.14}$$

In (4.14), $u(t)$ is the unit step function, $\omega_n = \sqrt{\frac{k}{m}}$, $\varsigma = \frac{c}{2\sqrt{km}}$, and $\omega_d = \omega_n\sqrt{1-\varsigma^2}$. The frequency transfer function is the Fourier transform of $h(t)$. It is given by

$$H(\omega) = \frac{1}{m} \frac{1}{\left(\omega_n^2 - \omega^2\right) + i2\varsigma\omega_n\omega}. \tag{4.15}$$

From (4.15), we have (4.16) to express $|H(w)|^2$ in the form

$$|H(\omega)|^2 = \frac{1}{m^2} \frac{1}{\left(\omega_n^2 - \omega^2\right)^2 + (2\varsigma\omega_n\omega)^2}. \tag{4.16}$$

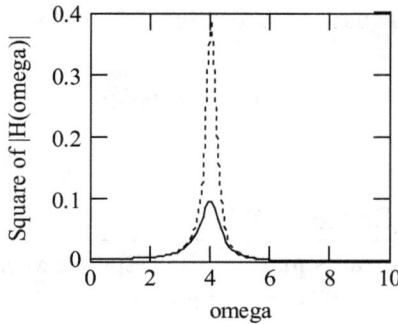

FIGURE 4.2 $|H(\omega)|^2$ with $\omega_n = 4$, $m = 1$, for $\zeta = 0.10$ (solid), 0.05 (dot).

Figure 4.2 indicates two plots of $|H(\omega)|^2$.

The Pierson and Moskowitz (P-M) spectrum is used to describe ocean surface waves. It is in the form

$$S_{\mathrm{PM}}(\omega) = ag^2\omega^{-5}e^{-b\left(\frac{g}{V}\right)^4\frac{1}{\omega^4}}. \qquad (4.17)$$

In (4.17), $a = 8.1\times10^{-3}$, $b = 0.74$, g is the acceleration of gravity (m/s²) and V wind speed (m/s) at an elevation of 19.5 m above the sea surface (Massel [2, p. 79]; also see Pierson and Moskowitz [3], and Li [4]). The unit of S is m²·s. In the example, we set $V = 15$. Thus, the driven PSD is

$$S_{pp}(\omega) = ag^2\omega^{-5}e^{-b\left(\frac{g}{15}\right)^4\frac{1}{\omega^4}}.$$

According to (4.6), we have the response PSD expressed by (4.18).

$$S_{xx}(\omega) = \frac{1}{m^2}\frac{ag^2\omega^{-5}e^{-b\left(\frac{g}{15}\right)^4\frac{1}{\omega^4}}}{\left(\omega_n^2 - \omega^2\right)^2 + (2\zeta\omega_n\omega)^2}. \qquad (4.18)$$

Figures 4.3 and 4.4 illustrate two plots of $S_{pp}(\omega)$ and $S_{xx}(\omega)$, respectively. The zero-frequency response $S_{xx}(0) = 0$ and the mean square response $r_{xx}(0) < \infty$.

Following Li [5–9], we have the corresponding driven and response signals as indicated in Figure 4.5.

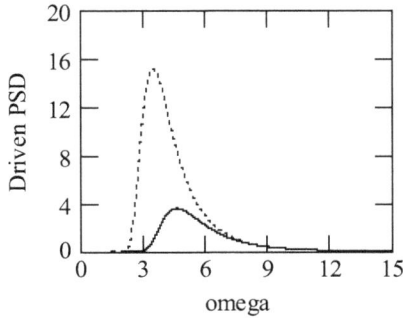

FIGURE 4.3   Plot of driven PSD $S_{pp}(\omega)$ for $V = 15$ (solid), 20 (dot).

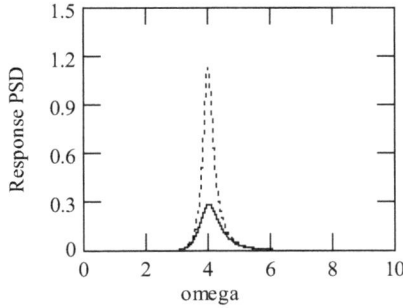

FIGURE 4.4   Plots of response PSD $S_{xx}(\omega)$ with $\omega_n = 4$, $m = 1$ when $\zeta = 0.10$ (solid), 0.05 (dot).

4.7 CROSS-CORRELATION RESPONSE AND CROSS-CORRELATION-BASED INPUT-OUTPUT RELATION

Based on (4.2), we consider $x(t + \tau)$ that is expressed by (4.19).

$$x(t+\tau) = \int_0^\infty h(u)p(t+\tau-u)du. \tag{4.19}$$

Taking into account the superposition, from (4.19), one has (4.20) to express $p(t)x(t + \tau)$

$$p(t)x(t+\tau) = \int_0^\infty h(u)p(t)p(t+\tau-u)du. \tag{4.20}$$

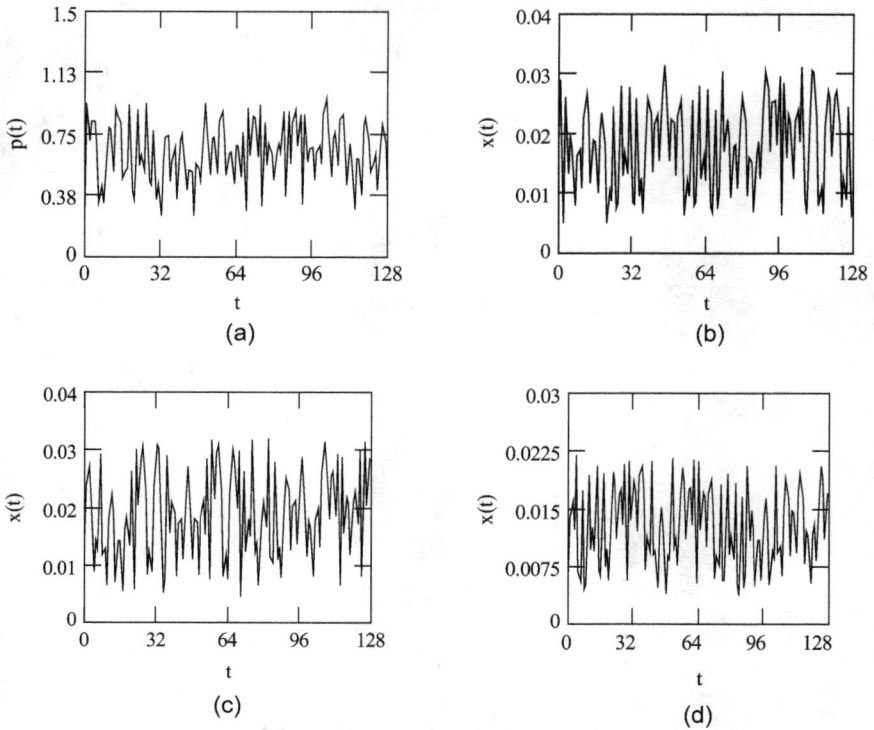

FIGURE 4.5 Plots of driven signal and response one. (a). Driven signal $p(t)$. (b). Response $x(t)$ with $\omega_n = 4$, $m = 1$ and $\zeta = 0.1$. (c). Response $x(t)$ with $\omega_n = 4$, $m = 1$ and $\zeta = 0.05$. (d). Response $x(t)$ with $\omega_n = 4$, $m = 1$ and $\zeta = 0.5$.

Therefore,

$$E\big[p(t)x(t+\tau)\big] = r_{px}(\tau) = \int_0^\infty h(u)E\big[p(t)p(t+\tau-u)\big]du$$

$$= \int_0^\infty h(u)r_{pp}(\tau-u)du. \tag{4.21}$$

Eq. (4.21) exhibits a relationship between the driven ACF $r_{pp}(\tau)$ and the cross-correlation response $r_{px}(\tau)$. To be precise,

$$r_{px}(\tau) = h(\tau) * r_{pp}(\tau). \tag{4.22}$$

Eq. (4.22) is the cross-correlation-based input-output relation.

4.8 CROSS-PSD RESPONSE AND CROSS-PSD-BASED INPUT-OUTPUT RELATION

Doing the Fourier transform on both sides of (4.22) results in

$$S_{px}(\omega) = S_{pp}(\omega)H(\omega). \tag{4.23}$$

Eq. (4.23) is the cross-PSD-based input-output relation. Generally, $S_{px}(w)$ is complex. Thus, (4.23) can be written by

$$\left|S_{px}(\omega)\right|e^{-i\theta_{px}(\omega)} = \left|H(\omega)\right|e^{-i\varphi(\omega)}S_{pp}(\omega). \tag{4.24}$$

Eq. (4.24) implies the expressions of (4.25) and (4.26).

$$\left|S_{px}(\omega)\right| = \left|H(\omega)\right|S_{pp}(\omega), \tag{4.25}$$

$$\theta_{px}(\omega) = \varphi(\omega). \tag{4.26}$$

The relation between $\left|S_{px}(\omega)\right|$ and $S_{xx}(\omega)$ can also be expressed by (4.27).

$$\left|S_{px}(\omega)\right| = \frac{S_{xx}(\omega)}{\left|H(\omega)\right|}. \tag{4.27}$$

4.9 COHERENCE FUNCTION

Multiplying the left side of (4.25) by the left side of (4.27) and multiplying the right side of (4.25) by the right side of (4.27) produce (4.28) to express $|S_{px}(\omega)|^2$ in the form

$$\left|S_{px}(\omega)\right|^2 = S_{pp}(\omega)S_{xx}(\omega). \tag{4.28}$$

From (4.28), we have an interesting expression in the form of (4.29).

$$\frac{\left|S_{px}(\omega)\right|^2}{S_{pp}(\omega)S_{xx}(\omega)} = 1. \tag{4.29}$$

The left side of (4.29) is a function of w. It is defined as a coherence function denoted by $\gamma_{px}^2(\omega)$. That is,

$$\gamma_{px}^2(\omega) = \frac{\left|S_{px}(\omega)\right|^2}{S_{pp}(\omega)S_{xx}(\omega)}. \tag{4.30}$$

It characterizes the coherence degree between the input $p(t)$ and the response $x(t)$ in the frequency domain. It satisfies

$$0 \leq \gamma_{px}^2(\omega) \leq 1. \tag{4.31}$$

Although

$$\gamma_{px}(\omega) = \frac{|S_{px}(\omega)|}{\sqrt{S_{pp}(\omega)S_{xx}(\omega)}}$$

and

$$0 \leq \gamma_{px}(\omega) \leq 1,$$

the term of coherence function is for $\gamma_{px}^2(\omega)$.

4.10 SUMMARY

We have described the responses of a random vibration system. They are PSD response, ACF response, cross PSD response, cross-correlation response, zero-frequency response, and mean square one. Input-output relations (PSD based, ACF based, cross PSD based, and cross-correlation based) have been explained. In addition, coherence function is described.

4.11 EXERCISES

4.1. Suppose that the PSD of a driven signal $p(t)$ follows the Pierson and Moskowitz (P-M) spectrum in the form

$$S_{pp}(\omega) = \alpha g^2 \omega^{-5} e^{-\beta \left(\frac{g}{V}\right)^4 \frac{1}{\omega^4}},$$

where $\alpha = 8.1 \times 10^{-3}$, $\beta = 0.74$, g is the acceleration of gravity (m/s²), and V wind speed (m/s) at an elevation of 19.5 m above the sea surface. Find the inverse Fourier transform of $S_{pp}(\omega)$.

4.2. Let the impulse response function $h(t)$ of a vibration system be

$$h(t) = \frac{1}{m\omega_d} e^{-\zeta\omega_n t} \sin\omega_d t u(t),$$

where $u(t)$ is the unit step function. Let $r_{pp}(t)$ be the inverse Fourier transform of $S_{pp}(\omega)$ given in Exercise 4.1. Find the cross-correlation response $r_{px}(t) = r_{pp}(t) * h(t)$.

4.3. In Exercise 4.2, find the ACF response $r_{xx}(t) = r_{pp}(t) * h(t) * h(^-t)$.

4.4. Let the ACF of an excitation be

$$r_{pp}(\tau) = e^{-2\lambda|\tau|}, \quad (\lambda > 0).$$

Let the impulse response function $h(t)$ of a vibration system be that in Exercise 4.2. Find the cross-correlation response $r_{px}(t) = r_{pp}(t) * h(t)$.

4.5. In Exercise 4.3, find the mean square response $r_{xx}(0)$.

4.6. In Exercise 4.2, find $F[r_{px}(t)] = S_{px}(\omega)$.

REFERENCES

1. E. A. Robinson, A historical perspective of spectrum estimation, *Proceedings of the IEEE*, vol. 70, no. 9, 1982, 885–907.
2. S. R. Massel, *Ocean Surface Waves: Their Physics and Prediction*, World Scientific, Singapore, 1997.
3. W. J. Pierson and L. Moskowitz, A proposed spectral form for fully developed wind seas based on the similarity theory of S. A. Kitaigorodskii, *Journal of Geophysical Research*, vol. 69, no. 24, 1964, 5181–5190.
4. M. Li, A method for requiring block size for spectrum measurement of ocean surface waves, *IEEE Transactions on Instrumentation and Measurement*, vol. 55, no. 6, 2006, 2207–2215.
5. M. Li, Experimental stability analysis of test system for doing fatigue test under random loading, *Journal of Testing and Evaluation*, vol. 34, no. 4, 2006, 364–367.
6. M. Li, An iteration method to adjusting random loading for a laboratory fatigue test, *International Journal of Fatigue*, vol. 27, no. 7, 2005, 783–789.
7. M. Li, An optimal controller of an irregular wave maker, *Applied Mathematical Modelling*, vol. 29, no. 1, 2005, 55–63.
8. M. Li, Generation of teletraffic of generalized Cauchy type, *Physica Scripta*, vol. 81, no. 2, 2010, 025007 (10pp).
9. M. Li, *Multi-Fractal Traffic and Anomaly Detection in Computer Communications*, CRC Press, Boca Raton, 2022.

CHAPTER 5

Vibrations with Frequency-Dependent Mass, Damping, and Stiffness

T HE HIGHLIGHTS IN THIS chapter are threefold. One is to show frequency-dependent elements (mass, damping, and stiffness) from the point of view of conventional structure vibrations. The second is to propose the general form of a vibration system with frequency-dependent elements (mass, damping, and stiffness). The third is to bring forward the closed-form analytic expressions about a vibration system with equivalent mass, damping, stiffness, damping ratio, natural angular frequencies, frequency ratio, responses (free, impulse, step), frequency transfer function, logarithmic decrement, and Q factor. This chapter is towards paving the way for fractional vibrations addressed in Chapters 6 and 7 in Volume I, and fractional random vibrations driven by fractional processes stated in Chapters 1–6 in Volume II.

5.1 INTRODUCTION

Conventionally, vibration elements, say, mass m, damping c, and stiffness k, are commonly assumed to be constants. In the conventional theory of vibrations, people pay attention to the phenomena of frequency-dependent

DOI: 10.1201/9781003657897-5

elements (mass or damping or stiffness), see, for example, Harris [1], Korotkin [2], Palley et al. [3], Kristiansen and Egeland [4], Zou et al. [5], Wu and Hsie [6], Qiao et al. [7], Jaberzadeh et al. [8], Xu et al. [9], Ghaemmaghami and Kwon [10], Hamdaoui et al. [11]. Since the theory of fractional vibrations established by Li [12–14] adopts frequency-dependent elements in the equivalent sense, we feel the usefulness of showing several realistic cases of frequency-dependent mass, damping, and stiffness respectively in Sections 5.2–5.4, so as to purposely write a general form of a vibration system with frequency-dependent mass, damping, and stiffness and discuss its vibration responses in Section 5.5. The intention of writing Section 5.5 is in two aspects. One is for the pavement of seven classes of fractional vibrators to be addressed in Chapters 6–7 in Volume I and Chapters 1–6 in Volume II. The other is to facilitate smoothing away possible hesitations as to why m and or c and or k may be frequency-dependent in fractional vibrations.

In this chapter, we describe frequency-dependent mass, damping, and stiffness, respectively in Sections 5.2–5.4. In Sections 5.5 and 5.6, we discuss a general vibration system with frequency-dependent elements. The concepts of equivalent logarithmic decrement and equivalent Q factor with respect to the general vibration system with frequency-dependent elements are given in Section 5.7. The summary is in Section 5.8.

5.2 CASES OF FREQUENCY-DEPENDENT MASS

5.2.1 Frequency-Dependent Mass in Auxiliary Mass Damper System

Consider a simple auxiliary mass damper indicated in Figure 5.1 (Harris [1, Chapter 6]). The system consists of a mass m_a, spring k_a, and viscous damper c_a.

Eq. (5.1) expresses the motion equation of the system with auxiliary mass in Figure 5.1

$$-k_a x_r(t) - c_a \frac{dx_r(t)}{dt} = m_a \frac{d^2[x_0(t) + x_r(t)]}{dt^2}. \tag{5.1}$$

Let X_r and X_0 be the phasors of $x_r(t)$ and $x_0(t)$, respectively. The phasor equation of the aforementioned is expressed by

$$\left(-k_a - i\omega c_a\right) X_r = -m_a \omega^2 (X_r + X_0). \tag{5.2}$$

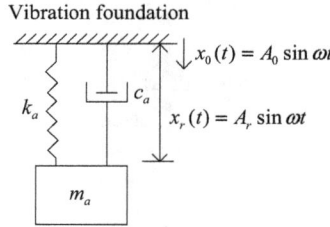

FIGURE 5.1   Auxiliary mass damper.

From (5.2), we have (5.3) to express the phasor X_r in the form

$$X_r = \frac{m_a \omega^2}{-m_a \omega^2 + k_a + i\omega c_a}.$$

(5.3)

Denote by F the phasor of the force exerting on the foundation. Then, (5.4) is its expression in the form

$$F = \frac{m_a \omega^2 \left(k_a + i\omega c_a \right)}{-m_a \omega^2 + k_a + i\omega c_a} X_0.$$

(5.4)

As the force acted by an equivalent mass m_{eq} is rigidly attached to the foundation, we have

$$F = m_{eq} \omega^2 X_0.$$

(5.5)

In (5.5), m_{eq} is given by

$$m_{eq} = \frac{k_a + i\omega c_a}{-m_a \omega^2 + k_a + i\omega c_a} m_a.$$

(5.6)

Rewriting (5.6) yields

$$m_{eq} = \frac{\left(k_a + i\omega c_a \right)\left(k_a - m_a \omega^2 - i\omega c_a \right)}{\left(k_a - m_a \omega^2 \right)^2 + \left(\omega c_a \right)^2} m_a = \frac{k_a \left(k_a - m_a \omega^2 \right) + \left(\omega c_a \right)^2 - i m_a c_a \omega^3}{\left(k_a - m_a \omega^2 \right)^2 + \left(\omega c_a \right)^2} m_a.$$

(5.7)

Eq. (5.6) or (5.7) exhibits that m_{eq} is complex. In the polar system, it may be expressed by

$$m_{eq} = |m_{eq}| \mathrm{Arg} m_{eq}.$$

(5.8)

In (5.8), $|m_{eq}|$ and $\mathrm{Arg}\,m_{eq}$ are expressed by (5.9) and (5.10), respectively,

$$|m_{eq}| = \frac{\left[k_a\left(k_a - m_a\omega^2\right) + \left(\omega c_a\right)^2\right]^2 + \left(m_a c_a \omega^3\right)^2}{\left(k_a - m_a\omega^2\right)^2 + \left(\omega c_a\right)^2}\, m_a, \tag{5.9}$$

$$\mathrm{Arg}\,m_{eq} = \tan^{-1}\frac{-i m_a c_a \omega^3}{k_a\left(k_a - m_a\omega^2\right) + \left(\omega c_a\right)^2}. \tag{5.10}$$

Note 5.1. An interesting point with respect to the structure in Figure 5.1 is that m_{eq} is complex. Both the modulus and argument of m_{eq} are the functions of ω.

Note 5.2. When $\omega = 0$, m_{eq} reduces to the primary mass m_a.

Note 5.3. In general, $0 \le |m_{eq}| < \infty$.

Note 5.4. When $c_a = 0$, m_{eq} is real.

Figure 5.2 illustrates a curve of $|m_{eq}|$.

5.2.2 Frequency-Dependent Mass in Added Mass

The frequency dependence of added mass is well known in the field of ship mechanics (Korotkin [2]). In general, a ship motion is with six degrees of freedom (Palley et al. [3]). We adopt the following symbols for discussions.

- q_n ($n = 1, \ldots, 6$): generalized coordinates.
- f_n: generalized forces.

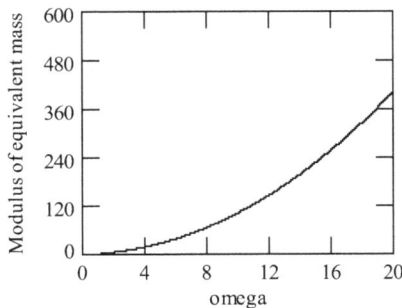

FIGURE 5.2 Illustration of $|m_{eq}|$ for $m_a = 1$, $c_a = 1$, and $k_a = 1$.

- m_{jn}: dry mass of the ship in direction j.

- c_{jn}: dry damping of the ship in direction j.

- k_{jn}: dry stiffness of the ship in direction j.

- $m_{\text{add}, jn}$: added mass of the ship in direction j.

- $h_{jn}(t)$: impulse response function in direction j to an impulse in velocity in direction n.

When $q_n(t) = q_n\cos(\omega t)$, according to Kristiansen and Egeland [4], one has an equation of motion in the form

$$\sum_{n=1}^{6}\left[m_{jn} + m_{\text{add}, jn}(\omega)\right]q_n'' + \sum_{n=1}^{6}c_{\text{eq}, jn}(\omega)q_n' + \sum_{n=1}^{6}k_{jn}q_n = f_j(t). \quad (5.11)$$

In (5.11), $f_j(t)$ is a sinusoidal force at ω, $m_{\text{add}, jn}(\omega)$ and $c_{jn}(\omega)$ are given by (5.12) and (5.13), respectively.

$$m_{\text{add}, jn}(\omega) = m_{jn} - \frac{1}{\omega}\int_0^{\infty} h_{jn}(t)\sin\omega t\, dt, \quad (5.12)$$

$$c_{\text{eq}, jn}(\omega) = c_{jn} + \int_0^{\infty} h_{jn}(t)\cos\omega t\, dt. \quad (5.13)$$

We write (5.14) to express the equivalent mass m_{eq} in the form

$$m_{\text{eq}} = m_{jn} + m_{\text{add}, jn}(\omega). \quad (5.14)$$

Therefore, we see that the equivalent mass m_{eq} of a ship in general is frequency dependent. Consequently, $m_{\text{eq}} = m_{\text{eq}}(\omega)$.

There are other types of expressions with respect to frequency-dependent mass in vibrations, see, for example, Zou et al. [5], Wu and Hsieh [6], Qiao et al. [7], Jaberzadeh et al. [8], Xu et al. [9], Ghaemmaghami and Kwon [10], Hamdaoui et al. [11], Li [12–14], Banerjee [15], White et al. [16], Dumont and Oliveira [17], Zhang et al. [18], Sun et al. [19].

5.3 CASES OF FREQUENCY-DEPENDENT DAMPING

5.3.1 Rigidly Connected Coulomb Damper

Have a look at Figure 5.3, which indicates a rigidly connected Coulomb damper.

The motion equation is expressed by (5.15) in the form

$$mx'' + k(x-u) \pm F_f = F_0 + \sin \omega t. \tag{5.15}$$

Since there is discontinuity in the damping force that occurs as the sign of the velocity changes at each half cycle, a step-by-step solution of the previous statement is required (Harris [1], Den Hartog [20]). Let $\delta = x - u$. Using the equivalence of energy dissipation for equating the energy dissipation per cycle for viscous-damped and Coulomb-damped systems produces (Harris [1], Jacobsen [21]) results in

$$\pi c_{eq} \omega \delta_0^2 = 4 F_f \delta_0. \tag{5.16}$$

In (5.16), the left side refers to the viscous-damped system and the right side to the Coulomb-damped system. The symbol δ_0 is the amplitude of relative displacement across the damper.

From (5.16), one has the equivalent damping coefficient for a Coulomb-damped system that has equivalent energy dissipation in the form

$$c_{eq} = \frac{4 F_f}{\pi \omega \delta_0}. \tag{5.17}$$

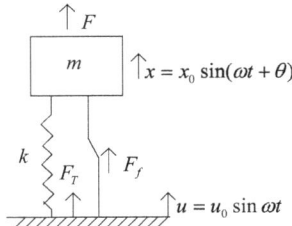

FIGURE 5.3   Rigidly connected Coulomb damper.

Eq. (5.17) exhibits that c_{eq} is frequency dependent. Hence, (5.18) is valid.

$$c_{eq} = c_{eq}(\omega). \tag{5.18}$$

5.3.2 Rayleigh Damping Assumption

The Rayleigh damping introduced by Rayleigh [22] is widely adopted in the field, see, for example, Harris [1], Palley et al. [3], Li [12–14], Jin and Xia [23], Trombetti and Silvestri [24, 25], Mohammad et al. [26], and Kim and Wiebe [27]. The standard form of the Rayleigh damping assumption is given by

$$c_{Raylegh} = am + bk. \tag{5.19}$$

In (5.19), a is proportional to ω while b is inversely proportional to ω. Thus, we may write

$$c_{Raylegh} = c_{Raylegh}(\omega). \tag{5.20}$$

Eq. (5.19) or (5.20) exhibits that the frequency dependence is a radical property of the damping Rayleigh assumed.

5.3.3 Remarks

Eq. (5.13) exhibits a case of frequency-dependent damping in ship motion, (5.17) shows a case of frequency-dependent damping for the structure in Figure 5.3, and (5.19) is about the Rayleigh damping. Other types of frequency dependent dampers in vibrations may refer to Kuo et al. [28], Stollwitzer et al. [29], Jith and Sarkar [30], Zhou et al. [31], Zarraga et al. [32], Xie et al. [33, 34], Hu et al. [35], Rouleau et al. [36], Hamdaoui et al. [37], Deng et al. [38], Dai et al. [39], Adessina et al. [40], Chang et al. [41], Lin et al. [42], Dai et al. [43], Catania and Sorrentino [44, 45], Zhang and Turner [46], Yoshida et al. [47], Assimaki and Kausel [48], Pan et al. [49], Ghosh and Viswanath [50], Mcdaniel et al. [51], Zhang et al. [52], Wang et al. [53], Lundén and Dahlberg [54], Figueroa et al. [55], Lázaro [56], and Crandall [57], simply citing a few.

5.4 CASES OF FREQUENCY-DEPENDENT STIFFNESS

5.4.1 Frequency-Dependent Stiffness in a Shaft-Driven by a Periodic Force

Consider a shaft driven by a periodic force, as shown in Figure 5.4. The mass m is supported by two springs with the primary stiffness k. Under the excitation of a force in axis direction, there is a force produced by displacement in the form $\frac{x}{l} F \cos \omega t$.

Thus, one may use (5.21) to express its motion equation in the form

$$mx'' + kx - \frac{x}{l} F \cos \omega t = 0. \tag{5.21}$$

Eq. (5.21) is actually in the type of the Mathieu's equation. Denote by k_{eq} the equivalent stiffness of the system. When writing (5.21) by (5.22)

$$mx'' + k_{eq}x = 0, \tag{5.22}$$

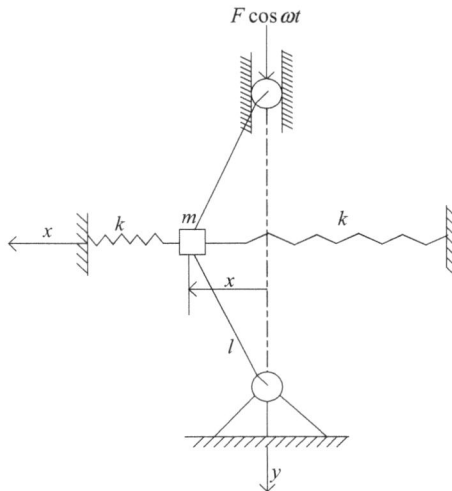

FIGURE 5.4   A shaft excited by a periodic force.

we have k_{eq} in the form

$$k_{eq} = k - \frac{F \cos \omega t}{l}. \tag{5.23}$$

Eq. (5.23) clearly designates that the equivalent stiffness k_{eq} is frequency dependent. Hence, $k_{eq} = k_{eq}(\omega)$.

5.4.2 Frequency-Dependent Stiffness: Simple Pendulum Case

Let l be the length of a simple pendulum. Denote by m the mass of the simple pendulum. Suppose that the fulcrum position of the pendulum moves periodically as $A_0 \cos xl$, see Figure 5.5.

The motion equation of the simple pendulum is given by

$$ml\theta'' + m\left(g - \omega^2 A_0 \cos \omega t\right) \sin \theta = 0. \tag{5.24}$$

In (5.24), when θ is small such that $\sin \theta \approx \theta$, we have

$$ml\theta'' + m\left(g - \omega^2 A_0 \cos \omega t\right)\theta = 0. \tag{5.25}$$

Eq. (5.25) is an approximation of (5.24). Replacing θ by x in (5.25) yields

$$mx'' + \frac{m}{l}\left(g - \omega^2 A_0 \cos \omega t\right)x = 0. \tag{5.26}$$

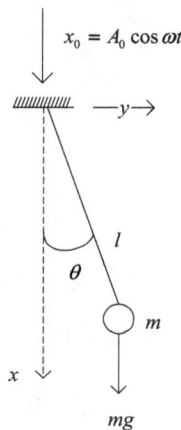

FIGURE 5.5   A simple pendulum.

In (5.26), we denote by k_{eq} the equivalent stiffness. Then,

$$k_{eq} = \frac{m}{l}\left(g - \omega^2 A_0 \cos\omega t\right).$$ (5.27)

Therefore, the motion equation (5.26) is expressed by (5.28).

$$mx'' + k_{eq}x = 0.$$ (5.28)

Eq. (5.27) exhibits that the stiffness k_{eq} is frequency dependent.

The topic of frequency-dependent stiffness attracts the interests of researchers. The other references regarding frequency-dependent stiffness in vibrations refer to Li [12–14], Banerjee [15], White et al. [16], Dumont and de Oliveira [17], Zhang et al. [18], Sun et al. [19], Yoshida et al. [47], Wu et al. [58], Blom and Kari [59], Gao et al. [60], Song et al. [61], Liu et al. [62], Zhang et al. [63], Banerjee et al. [64, 65], Lu et al. [66], Sung et al. [67], Mezghani et al. [68], Liu et al. [69], Kong et al. [70], Ege et al. [71], Mukhopadhyay et al. [72], Sainz-AjaIsidro et al. [73], Bozyigit [74], Varghese et al. [75], Failla et al. [76], Fan et al. [77], Roozen et al. [78], Mochida and Ilanko [79], just citing a few.

5.5 GENERAL VIBRATION SYSTEM WITH FREQUENCY-DEPENDENT ELEMENTS

5.5.1 Motion Equation

Based on the previous discussions, we write the general form of motion equation with frequency-dependent elements in the form

$$m_{eq}(\omega)x'' + c_{eq}(\omega)x' + k_{eq}(\omega)x = f(t),$$ (5.29)

where $f(t)$ is an excitation force. In other words, (5.29) stands for an equivalent motion equation.

Let $X(\omega)$ and $F(\omega)$ be the Fourier transform of $x(t)$ and $f(t)$, respectively. Then, the motion equation in the frequency domain is expressed by (5.30) in the form

$$\left[-\omega^2 m_{eq}(\omega) + i\omega c_{eq}(\omega) + k_{eq}(\omega)\right]X(\omega) = F(\omega).$$ (5.30)

5.5.2 Vibration Parameters

Denote by ω_{eqn} the equivalent damping-free natural angular frequency. It is given by

$$\omega_{eqn} = \sqrt{\frac{k_{eq}(\omega)}{m_{eq}(\omega)}}. \tag{5.31}$$

Since either m_{eq} or k_{eq} is a function of ω in (5.31), ω_{eqn} is a function of ω. Thus, the representation of (5.32) is sound.

$$\omega_{eqn} = \omega_{eqn}(\omega). \tag{5.32}$$

Let $\varsigma_{eq}(\omega)$ be the equivalent damping ratio. Define it by

$$\varsigma_{eq}(\omega) = \frac{1}{2} \frac{c_{eq}}{\sqrt{m_{eq}k_{eq}}}. \tag{5.33}$$

Using (5.31) and (5.33), therefore, we rewrite (5.29) by (5.34) in the form

$$x'' + 2\varsigma_{eq}(\omega)\omega_{eqn}(\omega)x' + \omega_{eqn}^2(\omega)x = \frac{f(t)}{m_{eq}(\omega)}. \tag{5.34}$$

Denote by $\omega_{eqd}(\omega)$ the equivalent damped natural angular frequency. Since $|\varsigma_{eq}(\omega)| > 1$ does not make sense in vibrations (Harris [1], Palley et al. [2], Li [13], Nakagawa and Ringo [80]), we restrict ς_{eq} by $|\varsigma_{eq}(\omega)| \leq 1$. Thus, (5.35) is its expression

$$\omega_{eqd}(\omega) = \omega_{eqn}(\omega)\sqrt{1 - \varsigma_{eq}^2(\omega)}. \tag{5.35}$$

Eq. (5.36) is the equivalent frequency ratio

$$\gamma_{eq} = \gamma_{eq}(\omega) = \frac{\omega}{\omega_{eqn}(\omega)}. \tag{5.36}$$

5.5.3 Free Response of General Vibration System with Frequency-Dependent Elements

The free response to a general vibration system with frequency-dependent elements is the solution to the equation given by

$$\begin{cases} m_{eq}(\omega)x''(t) + c_{eq}(\omega)x'(t) + k_{eq}(\omega)x(t) = 0, \\ x(0) = x_0, x'(0) = v_0. \end{cases} \qquad (5.37)$$

Rewrite (5.37) by

$$\begin{cases} x'' + 2\varsigma_{eq}(\omega)\omega_{eqn}(\omega)x' + \omega_{eqn}^2(\omega)x = 0, \\ x(0) = x_0, x'(0) = v_0. \end{cases} \qquad (5.38)$$

Then, (5.39) is the solution to (5.38). It is in the form

$$x(t) = e^{-\varsigma_{eq}\omega_{eqn}t}\left(x_0 \cos\omega_{eqd}t + \frac{v_0 + \varsigma_{eq}\omega_{eqn}x_0}{\omega_{eqd}}\sin\omega_{eqd}t\right), \quad t \geq 0. \quad (5.39)$$

5.5.4 Impulse Response of General Vibration System with Frequency-Dependent Elements

When investigating the impulse response to a general vibration system with frequency-dependent elements, we use (5.40).

$$\begin{cases} h''(t) + 2\varsigma_{eq}(\omega)\omega_{eqn}(\omega)h'(t) + \omega_{eqn}^2(\omega)h(t) = \dfrac{\delta(t)}{m_{eq}(\omega)}, \\ h(0) = 0, h'(0) = 0. \end{cases} \qquad (5.40)$$

Eq. (5.41) is the solution to (5.40). It is expressed by

$$h(t) = e^{-\varsigma_{eq}\omega_{eqn}t}\frac{1}{m_{eq}\omega_{eqd}}\sin\omega_{eqd}t, \quad t \geq 0. \qquad (5.41)$$

5.5.5 Step Response of General Vibration System with Frequency-Dependent Elements

Denote by $g(t)$ the step response to a general vibration system with frequency-dependent elements. Considering (5.42)

$$\begin{cases} g''(t) + 2\varsigma_{eq}(\omega)\omega_{eqn}(\omega)g'(t) + \omega_{eqn}^2(\omega)g(t) = \dfrac{u(t)}{m_{eq}(\omega)}, \\ g(0) = 0, g'(0) = 0, \end{cases} \qquad (5.42)$$

we have (5.43) with (5.44) as the solution to (5.42). That is,

$$g(t) = \frac{1}{k_{eq}(\omega)}\left[1 - \frac{e^{-\varsigma_{eq}\omega_{eqn}t}}{\sqrt{1-\varsigma_{eq}^2}}\cos\left(\omega_{eqd}t-\phi\right)\right], \quad t \geq 0, \qquad (5.43)$$

where

$$\phi = \tan^{-1}\frac{\varsigma_{eq}}{\sqrt{1-\varsigma_{eq}^2}}. \qquad (5.44)$$

5.6 FREQUENCY TRANSFER FUNCTION OF GENERAL VIBRATION SYSTEM WITH FREQUENCY-DEPENDENT ELEMENTS

Let $H(\omega)$ be the Fourier transform of $h(t)$. From (5.40), we have

$$\left[\omega_{eqn}^2(\omega) - \omega^2 + i2\varsigma_{eq}(\omega)\omega_{eqn}(\omega)\omega\right]H(\omega) = \frac{1}{m_{eq}(\omega)}. \qquad (5.45)$$

From (5.45), we have (5.46) to express $H(\omega)$ in the form

$$H(\omega) = \frac{1}{m_{eq}(\omega)\left[\omega_{eqn}^2(\omega) - \omega^2 + i2\varsigma_{eq}(\omega)\omega_{eqn}(\omega)\omega\right]}$$
$$= \frac{1}{k_{eq}(\omega)\left[1 - \gamma_{eq}^2 + i2\varsigma_{eq}(\omega)\gamma_{eq}\right]}. \qquad (5.46)$$

The amplitude $|H(\omega)|$ is given by (5.47).

$$|H(\omega)| = \frac{1/k_{eq}}{\sqrt{\left(1-\gamma_{eq}^2\right)^2 + \left(2\varsigma_{eq}\gamma_{eq}\right)^2}}. \qquad (5.47)$$

The phase is expressed by (5.48) in the form

$$\varphi(\omega) = -\tan^{-1}\frac{2\varsigma_{eq}(\omega)\gamma_{eq}}{1-\gamma_{eq}^2}. \qquad (5.48)$$

When computing $\phi(\omega)$ using digital computers, we may use (5.49)

$$\varphi(\omega) = \cos^{-1} \frac{1 - \gamma_{eq}^2}{\sqrt{\left(1 - \gamma_{eq}^2\right)^2 + \left(2\varsigma_{eq}\gamma_{eq}\right)^2}}. \tag{5.49}$$

5.7 LOGARITHMIC DECREMENT AND Q FACTOR OF GENERAL VIBRATION SYSTEM WITH FREQUENCY-DEPENDENT ELEMENTS

Let t_i and t_{i+1} be two time points of the free response $x(t)$, where $x(t_i)$ and $x(t_{i+1})$ are successive peak values at t_i and t_{i+1}. Let Δ_{eq} be the equivalent logarithmic decrement of $x(t)$. Then, (5.50) is its expression. It is in the form

$$\Delta_{eq} = \Delta_{eq}(\omega) = \ln \frac{x(t_i)}{x(t_{i+1})} = \frac{2\pi\varsigma_{eq}(\omega)}{\sqrt{1 - \varsigma_{eq}^2(\omega)}}. \tag{5.50}$$

Let Q_{eq} be the equivalent Q factor of a general vibration system with frequency-dependent elements. Then, (5.51) is its expression. It is given by

$$Q_{eq} = Q_{eq}(\omega) = \frac{1}{2\varsigma_{eq}(\omega)}. \tag{5.51}$$

5.8 SUMMARY

We have described several cases of structures that have frequency-dependent mass, damping, or stiffness. The general vibration system with frequency-dependent elements is analysed by representing its equivalent motion equation, equivalent vibration parameters (damping ratio, natural angular frequencies, frequency ratio), responses (free, impulse, step), frequency transfer function, equivalent logarithmic decrement, and equivalent Q factor.

5.9 EXERCISES

5.1. In (5.23), find the condition for $k_{eq} < 0$.

5.2. In (5.27), find the condition for $k_{eq} < 0$.

5.3. Let

$$mx'' + k_{eq}x = 0,$$

where

$$k_{eq} = \frac{m}{l}\left(g - \omega^2 A_0 \cos\omega t\right).$$

Find the free response when $x(0) = 1$ and $x'(0) = 0$.

5.4. Suppose that a vibration system is with the primary mass m and stiffness k. Its motion equation is expressed by

$$-m\omega^{\alpha-2}\cos\frac{\alpha\pi}{2}\frac{d^2x(t)}{dt^2} + m\omega^{\alpha-1}\sin\frac{\alpha\pi}{2}\frac{dx(t)}{dt} + kx(t) = 0, \quad 1 < \alpha < 3.$$

Write the equivalent mass and damping of the system discussed.

5.5. For the vibration system in Exercise 5.4, find its free response for $x(0) = 1$ and $x'(0) = 0$.

5.6. For the vibration system in Exercise 5.4, find its impulse response.

5.7. For the vibration system in Exercise 5.4, find its unit step response.

5.8. For the vibration system in Exercise 5.4, find its sinusoidal response for $x(0) = 0$ and $x'(0) = 0$.

REFERENCES

1. C. M. Harris, *Shock and Vibration Handbook*, 5th Ed., McGraw-Hill, New York, 2002.
2. A. I. Korotkin, *Added Masses of Ship Structures, Fluid Mechanics and Its Applications*, vol. 88, Springer, Netherlands, 2009.
3. O. M. Palley, Г. B. Bahizov, and E. Я. Voroneysk, *Handbook of Ship Structural Mechanics*, National Defense Industry Publishing House, Beijing, China, 2002. In Chinese. Translated from Russian by B. H. Xu, X. Xu, and M. Q. Xu.
4. E. Kristiansen and O. Egeland, Frequency-dependent added mass in models for controller design for wave motion damping, *IFAC Proceedings Volumes*, vol. 36, no. 21, 2003, 67–72.

5. M.-S. Zou, Y.-S. Wu, Y.-M. Liu, and C.-G. Lin, A three-dimensional hydro-elasticity theory for ship structures in acoustic field of shallow sea, *Journal of Hydrodynamics*, vol. 25, 2013, 929–937.

6. J.-S. Wu and M. Hsieh, An experimental method for determining the frequency-dependent added mass and added mass moment of inertia for a floating body in heave and pitch motions, *Ocean Engineering*, vol. 28, no. 4, 2001, 417–438.

7. Y. Qiao, J. Zhang, and P. Zhai, A magnetic field- and frequency-dependent dynamic shear modulus model for isotropic silicone rubber-based magnetorheological elastomers, *Composites Science and Technology*, vol. 204, 2021, 108637.

8. M. Jaberzadeh, B. Li, and K. T. Tan, Wave propagation in an elastic metamaterial with anisotropic effective mass density, *Wave Motion*, vol. 89, 2019, 131–141.

9. C. Xu, M.-Z. Wu, and M. Hamdaoui, Mixed integer multi-objective optimization of composite structures with frequency-dependent interleaved viscoelastic damping layers, *Computers & Structures*, vol. 172, 2016, 81–92.

10. A. R. Ghaemmaghami and O.-S. Kwon, Nonlinear modeling of MDOF structures equipped with viscoelastic dampers with strain, temperature and frequency-dependent properties, *Engineering Structures*, vol. 168, 2018, 903–914.

11. M. Hamdaoui, G. Robin, M. Jrad, and E. M. Daya, Optimal design of frequency dependent three-layered rectangular composite beams for low mass and high damping, *Composite Structures*, vol. 120, 2015, 174–182.

12. M. Li, Three classes of fractional oscillators, *Symmetry-Basel*, vol. 10, no. 2, 2018 (91 pages).

13. M. Li, *Fractional Vibrations with Applications to Euler-Bernoulli Beams*, CRC Press, Boca Raton, 2023.

14. M. Li, Dealing with stationary sinusoidal responses of seven types of multi-fractional vibrators using multi-fractional phasor, *Symmetry*, vol. 16, no. 9, 2024, 1197.

15. J. R. Banerjee, Frequency dependent mass and stiffness matrices of bar and beam elements and their equivalency with the dynamic stiffness matrix, *Computers & Structures*, vol. 254, 2021, 106616.

16. R. E. White, J. H. G. Macdonald, and N. A. Alexander, A nonlinear frequency-dependent spring-mass model for estimating loading caused by rhythmic human jumping, *Engineering Structures*, vol. 240, 2021, 112229.

17. N. A. Dumont and R. de Oliveira, From frequency-dependent mass and stiffness matrices to the dynamic response of elastic systems, *International Journal of Solids and Structures*, vol. 38, no. 10–13, 2001, 1813–1830.

18. J. Zhang, D. Yao, M. Shen, X. Sheng, J. Li, and S. Guo, Temperature- and frequency-dependent vibroacoustic response of aluminium extrusions damped with viscoelastic materials, *Composite Structures*, vol. 272, 2021, 114148.

19. P. Sun, H. Yang, and Y. Zhao, Time-domain calculation method of improved hysteretic damped system based on frequency-dependent loss factor, *Journal of Sound and Vibration*, vol. 488, 2020, 115658.

20. J. P. Den Hartog, *Mechanical Vibrations*, 4th Ed., McGraw-Hill, New York, 1956.

21. L. S. Jacobsen, Steady forced vibrations as influenced by damping, *Transactions of the American Society of Mechanical Engineers*, vol. 52, no. 1, 1930, 169–181.

22. J. W. Strutt, 3rd Baron Rayleigh, M. A., F. R. S., *The Theory of Sound*, vol. 1, Macmillan & Co., Ltd., London, 1877.

23. X. D. Jin and L. J. Xia, *Ship Hull Vibration*, The Press of Shanghai Jiaotong University, Shanghai, China, 2011. In Chinese.

24. T. Trombetti and S. Silvestri, On the modal damping ratios of shear-type structures equipped with Rayleigh damping systems, *Journal of Sound and Vibration*, vol. 292, no. 1–2, 2006, 21–58.

25. T. Trombetti and S. Silvestri, Novel schemes for inserting seismic dampers in shear-type systems based upon the mass proportional component of the Rayleigh damping matrix, *Journal of Sound and Vibration*, vol. 302, no. 3, 2007, 486–526.

26. D. R. A. Mohammad, N. U. Khan, and V. Ramamurti, On the role of Rayleigh damping, *Journal of Sound and Vibration*, vol. 185, no. 2, 1995, 207–218.

27. H.-G. Kim and R. Wiebe, Experimental and numerical investigation of nonlinear dynamics and snap-through boundaries of post-buckled laminated composite plates, *Journal of Sound and Vibration*, vol. 439, 2019, 362–387.

28. C.-H. Kuo, J.-Y. Huang, C.-M. Lin, C.-T. Chen, and K.-L. Wen, Near-surface frequency-dependent nonlinear damping ratio observation of ground motions using SMART1, *Soil Dynamics and Earthquake Engineering*, vol. 147, 2021, 106798.

29. A. Stollwitzer, J. Fink, and T. Malik, Experimental analysis of damping mechanisms in ballasted track on single-track railway bridges, *Engineering Structures*, vol. 220, 2020, 110982.

30. J. Jith and S. Sarkar, A model order reduction technique for systems with nonlinear frequency dependent damping, *Applied Mathematical Modelling*, vol. 77, Part 2, 2020, 1662–1678.

31. Y. Zhou, A. Liu, and Y. Jia, Frequency-dependent orthotropic damping properties of Nomex honeycomb composites, *Thin-Walled Structures*, vol. 160, 2021, 107372.

32. O. Zarraga, I. Sarría, J. García-Barruetabeña, and F. Cortés, Dynamic analysis of plates with thick unconstrained layer damping, *Engineering Structures*, vol. 201, 2019, 109809.

33. X. Xie, H. Zheng, S. Jonckheere, and W. Desmet, Explicit and efficient topology optimization of frequency-dependent damping patches using moving morphable components and reduced-order models, *Computer Methods in Applied Mechanics and Engineering*, vol. 355, 2019, 591–613.

34. X. Xie, H. Zheng, S. Jonckheere, A. de Walle, B. Pluymers, and W. Desmet, Adaptive model reduction technique for large-scale dynamical systems with frequency-dependent damping, *Computer Methods in Applied Mechanics and Engineering*, vol. 332, 2018, 363–381.
35. J. Hu, J. Ren, Z. Zhe, M. Xue, Y. Tong, J. Zou, Q. Zheng, and H. Tang, A pressure, amplitude and frequency dependent hybrid damping mechanical model of flexible joint, *Journal of Sound and Vibration*, vol. 471, 2020, 115173.
36. L. Rouleau, J.-F. Deü, and A. Legay, A comparison of model reduction techniques based on modal projection for structures with frequency-dependent damping, *Mechanical Systems and Signal Processing*, vol. 90, 2017, 110–125.
37. M. Hamdaoui, K. S. Ledi, G. Robin, and E. M. Daya, Identification of frequency-dependent viscoelastic damped structures using an adjoint method, *Journal of Sound and Vibration*, vol. 453, 2019, 237–252.
38. Y. Deng, S. Y. Zhang, M. Zhang, and P. Gou, Frequency-dependent aerodynamic damping and its effects on dynamic responses of floating offshore wind turbines, *Ocean Engineering*, vol. 278, 2023, 114444.
39. X.-J. Dai, J.-H. Lin, H.-R. Chen, and F. W. Williams, Random vibration of composite structures with an attached frequency-dependent damping layer, *Composites Part B: Engineering*, vol. 39, no. 2, 2008, 405–413.
40. A. Adessina, M. Hamdaoui, C. Xu, and E. M. Daya, Damping properties of bi-dimensional sandwich structures with multi-layered frequency-dependent visco-elastic cores, *Composite Structures*, vol. 154, 2016, 334–343.
41. D.-W. Chang, J. M Roesset, and C.-H. Wen, A time-domain viscous damping model based on frequency-dependent damping ratios, *Soil Dynamics and Earthquake Engineering*, vol. 19, no. 8, 2000, 551–558.
42. T. R. Lin, N. H. Farag, and J. Pan, Evaluation of frequency dependent rubber mount stiffness and damping by impact test, *Applied Acoustics*, vol. 66, no. 7, 2005, 829–844.
43. Q. Dai, Z. Qin, and F. Chu, Parametric study of damping characteristics of rotating laminated composite cylindrical shells using Haar wavelets, *Thin-Walled Structures*, vol. 161, 2021, 107500.
44. G. Catania and S. Sorrentino, Dynamical analysis of fluid lines coupled to mechanical systems taking into account fluid frequency-dependent damping and non-conventional constitutive models: Part 1—Modeling fluid lines, *Mechanical Systems and Signal Processing*, vol. 50–51, 2015, 260–280.
45. G. Catania and S. Sorrentino, Dynamical analysis of fluid lines coupled to mechanical systems taking into account fluid frequency-dependent damping and non-conventional constitutive models: Part 2—Coupling with mechanical systems, *Mechanical Systems and Signal Processing*, vol. 50–51, 2015, 281–295.
46. W. Zhang and K. Turner, Frequency dependent fluid damping of micro/nano flexural resonators: Experiment, model and analysis, *Sensors and Actuators A: Physical*, vol. 134, no. 2, 2007, 594–599.

47. N. Yoshida, S. Kobayashi, and K. Miura, Equivalent linear method considering frequency dependent characteristics of stiffness and damping, *Soil Dynamics and Earthquake Engineering*, vol. 22, no. 3, 2002, 205–222.

48. D. Assimaki and E. Kausel, An equivalent linear algorithm with frequency- and pressure-dependent moduli and damping for the seismic analysis of deep sites, *Soil Dynamics and Earthquake Engineering*, vol. 22, no. 9–12, 2002, 959–965.

49. S. Pan, Q. Dai, Z. Qin, and F. Chu, Damping characteristics of carbon nanotube reinforced epoxy nanocomposite beams, *Thin-Walled Structures*, vol. 166, 2021, 108127.

50. M. K. Ghosh and N. S. Viswanath, Frequency dependent stiffness and damping coefficients of orifice compensated multi-recess hydrostatic journal bearings, *International Journal of Machine Tools and Manufacture*, vol. 27, no. 3, 1987, 275–287.

51. J. G. Mcdaniel, P. Dupont, and L. Salvino, A wave approach to estimating frequency-dependent damping under transient loading, *Journal of Sound and Vibration*, vol. 231, no. 2, 2000, 433–449.

52. H. Zhang, X. Ding, and H. Li, Topology optimization of composite material with high broadband damping, *Computers & Structures*, vol. 239, 2020, 106331.

53. X. Wang, X. Li, R.-P. Yu, J.-W. Ren, Q.-C. Zhang, Z.-Y. Zhao, C.-Y. Ni, B. Han, and T. J. Lu, Enhanced vibration and damping characteristics of novel corrugated sandwich panels with polyurea-metal laminate face sheets, *Composite Structures*, vol. 251, 2020, 112591.

54. R. Lundén and T. Dahlberg, Frequency-dependent damping in structural vibration analysis by use of complex series expansion of transfer functions and numerical Fourier transformation, *Journal of Sound and Vibration*, vol. 80, no. 2, 1982, 161–178.

55. A. Figueroa, M. Telenko, L. Chen, and S. F. Wu, Determining structural damping and vibroacoustic characteristics of a non-symmetrical vibrating plate in free boundary conditions using the modified Helmholtz equation least squares method, *Journal of Sound and Vibration*, vol. 495, 2021, 115903.

56. M. Lázaro, Critical damping in nonviscously damped linear systems, *Applied Mathematical Modelling*, vol. 65, 2019, 661–675.

57. S. H. Crandall, The role of damping in vibration theory, *Journal of Sound and Vibration*, vol. 11, no. 1, 1970, 3–18.

58. M. Y. Wu, H. Yin, X. B. Li, J. C. Lv, G. Q. Liang, and Y. T. Wei, A new dynamic stiffness model with hysteresis of air springs based on thermodynamics, *Journal of Sound and Vibration*, vol. 521, 2022, 116693.

59. P. Blom and L. Kari, The frequency, amplitude and magnetic field dependent torsional stiffness of a magneto-sensitive rubber bushing, *International Journal of Mechanical Sciences*, vol. 60, 2012, 54–58.

60. X. Gao, Q. Feng, A. Wang, X. Sheng, and G. Cheng, Testing research on frequency-dependent characteristics of dynamic stiffness and damping for high-speed railway fastener, *Engineering Failure Analysis*, vol. 129, 2021, 105689.

61. X. Song, H. Wu, H. Jin, and C. S. Cai, Noise contribution analysis of a U-shaped girder bridge with consideration of frequency dependent stiffness of rail fasteners, *Applied Acoustics*, vol. 205, 2023, 109280.

62. X. Liu, X. Zhao, S. Adhikari, and X. Liu, Stochastic dynamic stiffness for damped taut membranes, *Computers & Structures*, vol. 248, 2021, 106483.

63. X. Zhang, D. Thompson, H. Jeong, M. Toward, D. Herron, and N. Vincent, Measurements of the high frequency dynamic stiffness of railway ballast and subgrade, *Journal of Sound and Vibration*, vol. 468, 2020, 115081.

64. J. R. Banerjee, A. Ananthapuvirajah, and S. O. Papkov, Dynamic stiffness matrix of a conical bar using the Rayleigh-Love theory with applications, *European Journal of Mechanics—A/Solids*, vol. 86, 2021, 104144.

65. J. R. Banerjee, A. Ananthapuvirajah, X. Liu, and C. Sun, Coupled axial-bending dynamic stiffness matrix and its applications for a Timoshenko beam with mass and elastic axes eccentricity, *Thin-Walled Structures*, vol. 159, 2021, 107197.

66. T. Lu, A. V. Metrikine, and M. J. M. M. Steenbergen, The equivalent dynamic stiffness of a visco-elastic half-space in interaction with a periodically supported beam under a moving load, *European Journal of Mechanics—A/Solids*, vol. 84, 2020, 104065.

67. D. Sung, S. Chang, and S. Kim, Effect of additional anti-vibration sleeper track considering sleeper spacing and track support stiffness on reducing low-frequency vibrations, *Construction and Building Materials*, vol. 263, 2020, 120140.

68. F. Mezghani, A. F. del Rincón, M. A. B. Souf, P. G. Fernandez, F. Chaari, F. V. Rueda, and M. Haddar, Alternating frequency time domains identification technique: Parameters determination for nonlinear system from measured transmissibility data, *European Journal of Mechanics—A/Solids*, vol. 80, 2020, 103886.

69. X. Liu, D. Thompson, G. Squicciarini, M. Rissmann, P. Bouvet, G. Xie, J. Martínez-Casas, J. Carballeira, I. L. Arteaga, M. A. Garralaga, and J. A. Chover, Measurements and modelling of dynamic stiffness of a railway vehicle primary suspension element and its use in a structure-borne noise transmission model, *Applied Acoustics*, vol. 182, 2021, 108232.

70. X. Kong, X. Zeng, and K. Han, Dynamical measurements on viscoelastic behaviors of spiders in electro-dynamic loudspeakers, *Applied Acoustics*, vol. 104, 2016, 67–75.

71. K. Ege, N. B. Roozen, Q. Leclère, and R. G. Rinaldi, Assessment of the apparent bending stiffness and damping of multilayer plates; modelling and experiment, *Journal of Sound and Vibration*, vol. 426, 2018, 129–149.

72. T. Mukhopadhyay, S. Adhikari, and A. Alu, Probing the frequency-dependent elastic moduli of lattice materials, *Acta Materialia*, vol. 165, 2019, 654–665.

73. J. A. Sainz-AjaIsidro, A. Carrascal, and S. Diego, Influence of the operational conditions on static and dynamic stiffness of rail pads, *Mechanics of Materials*, vol. 148, 2020, 103505.

74. B. Bozyigit, Seismic response of pile supported frames using the combination of dynamic stiffness approach and Galerkin's method, *Engineering Structures*, vol. 244, 2021, 112822.

75. R. Varghese, A. Boominathan, and S. Banerjee, Stiffness and load sharing characteristics of piled raft foundations subjected to dynamic loads, *Soil Dynamics and Earthquake Engineering*, vol. 133, 2020, 106117.

76. G. Failla, R. Santoro, A. Burlon, and A. F. Russillo, An exact approach to the dynamics of locally-resonant beams, *Mechanics Research Communications*, vol. 103, 2020, 103460.

77. R.-L. Fan, Z.-N. Fei, B.-Y. Zhou, H.-B. Gong, and P.-J. Song, Two-step dynamics of a semiactive hydraulic engine mount with four-chamber and three-fluid-channel, *Journal of Sound and Vibration*, vol. 480, 2020, 115403.

78. N. B. Roozen, L. Labelle, Q. Leclère, K. Ege, and S. Alvarado, Non-contact experimental assessment of apparent dynamic stiffness of constrained-layer damping sandwich plates in a broad frequency range using a Nd: YAG pump laser and a laser doppler vibrometer, *Journal of Sound and Vibration*, vol. 395, 2017, 90–101.

79. Y. Mochida and S. Ilanko, On the Rayleigh-Ritz method, Gorman's super-position method and the exact dynamic stiffness method for vibration and stability analysis of continuous systems, *Thin-Walled Structures*, vol. 161, 2021, 107470.

80. K. Nakagawa and M. Ringo, *Engineering Vibrations*, Shanghai Science and Technology Publishing House, Shanghai, China, 1981. In Chinese. Translated from Japanese by S. R. Xia.

CHAPTER 6

Classification of Fractional Vibrations

T HIS CHAPTER GIVES THE classification of fractional vibration systems. We classify fractional vibrators into seven classes. A class I vibrator is damping free in form and only with fractional inertia force. A class II vibrator is only with fractional damping force. A class III vibration system is with both fractional inertia and fractional damping forces. A class IV system is damping free in form but with both fractional inertia and fractional restoration forces. A class V vibrator is damping free in form and only with fractional restoration force. A class VI vibration system is with fractional inertia, fractional damping, and fractional restoration forces. A class VII system is with both fractional damping and fractional restoration forces.

6.1 INTRODUCTION

Consider a vibration system with the primary mass m, primary damping c, and primary stiffness k. Conventionally, the forces of inertia, damping, and restoration are expressed by $mx''(t)$, $cx'(t)$, and $kx(t)$, respectively, where $x(t)$ is displacement. Although people study various structures for different expressions of the forces of inertia, damping, and restoration by introducing equivalent mass or damping or stiffness, see, for example, [1–19], and Chapter 5 in Volume I, acceleration, velocity, and displacement are defaulted to be expressed by $x''(t)$, $x'(t)$, and $x(t)$.

DOI: 10.1201/9781003657897-6

Elishakoff used to state a research direction of fractional vibrations from a view of model uncertainty [19, Chapter 14]. From the point of view of fractional vibrations, however, among three forces, namely, inertia, damping, and restoration, if one of them is of fractional order, we encounter the issue of fractional vibrations (Li [20], Luo [21]). Following Li [20], we list seven classes of fractional vibration systems in this chapter. The fractional calculus used in this research is of Weyl, citing [22–29] for fractional calculus.

6.2 CLASSIFICATION

6.2.1 Class I Fractional Vibrators

A class I fractional vibrator means that (6.1) is its motion equation

$$m\frac{d^\alpha x(t)}{dt^\alpha} + kx(t) = f(t), \quad 1 < \alpha < 3. \tag{6.1}$$

This class of vibrators only contains fractional inertia force $mx^{(\alpha)}(t)$. It is damping free in form; see for example, Li [20], Pskhu and Rekhviashvili [30, Eq. (1)], Rossikhin [31, Eq. (19b)], Čermák and Kisela [32, Eq. (2.1)], Duan et al. [33, Eq. (7)], Uchaikin [34], Uchaikin [35, Chapter 7], Li [36], Duan [37, Eq. (3)], Zurigat [38, Eq. (16)], Blaszczyk and Ciesielski [39, Eq. (1)], Blaszczyk et al. [40, Eq. (10)], Blaszczyk [41], Al-Rabtah et al. [42, Eq. (3.1)], Drozdov [43, Eq. (9)], Stanislavsky [44], Achar et al. ([45 Eq. (1)], [46, Eq. (9)], [47, Eq. (2)]), Tofighi [48, Eq. (2)], Ryabov and Puzenko [49, Eq. (1)], Ahmad and Elwakil [50, Eq. (1)], Duan et al. [51, Eq. (4.2)], Tavazoei [52], Sandev and Tomovski [53, Eq. (36)], and Singh et al. [54].

Conventionally, the range of α for vibrators in class I is $1 < \alpha \le 2$, see [30–54]. In our research, we make a development so that $1 < \alpha < 3$ (Li [20]).

6.2.2 Class II Fractional Vibrators

When the motion equation of a fractional vibrator is given by (6.2) for $0 < \beta < 2$

$$m\frac{d^2 x(t)}{dt^2} + c\frac{d^\beta x(t)}{dt^\beta} + kx(t) = f(t), \tag{6.2}$$

it is called a class II fractional vibrator (Li [20]).

The class II consists of vibrators only with fractional friction term $cx^{(\beta)}(t)$. The literature regarding the class II type vibrations is relatively rich, see, for example, Li [20] and [36], Luo [21, Chapter 1], Lin et al. [55, Eq. (2)], Duan [56, Eq. (31)], Alkhaldi et al. [57, Eq. (1a)], Dai et al. [58, Eq. (1)], Ren et al. [59, Eq. (1)], [60, Eq. (1)], Xu et al. [61, Eq. (1)], He et al. [62, Eq. (4)], Leung et al. [63, Eq. (2)], Chen et al. [64, Eq. (1)], Deü and Matignon [65, Eq. (1)], Drăgănescu et al. [66, Eq. (4)], Rossikhin and Shitikova [67, Eq. (3)], Xie and Lin [68, Eq. (1)], Yuan et al. [69, Eq. (8)], Nešić et al. [70, Eq. (3)], Lin et al. [71, Eq. (1)], Naranjani et al. [72, Eq. (1)], Duan and Zhang [73, Eq. (1)], Matteo et al. [74], Shitikova [75, Eq. (22)], Yıldız et al. [76, Eq. (25)], Hu et al. [77, Eq. (1a)], Cao et al. [78, Eq. (8)], [79, Eq. (8)], Wang et al. [80, Eq. (1)], Kaltenbacher and Schlintl [81, Eq. (7)], Spanos and Malara [82, Eq. (1)], [83, Eq. (7)], Tomovski and Sandev [84, Eq. (44)], Zelenev et al. [85, Eq. (2.4)], Rossikhin and Shitikova [86, Eq. (2.2.2)], [87, Eq. (26a)], Bagley and Torvik [88, Eq. (32)], Duan et al. [89, Eq. (7)], and Tsinker [90, Chapter 28].

Traditionally, the fractional order β is for $0 < \beta \leq 1$, see [35, 54–90]. However, we make a step further for $0 < \beta < 2$ in our research (Li [20]). Note that Luo [21, Chapter 1] also takes the range of $0 < \beta < 2$.

6.2.3 Class III Fractional Vibrators

A class III fractional vibrator implies that (6.3) stands for its vibration motion equation in the form

$$m\frac{d^\alpha x(t)}{dt^\alpha} + c\frac{d^\beta x(t)}{dt^\beta} + kx(t) = f(t), \qquad (6.3)$$

where $1 < \alpha < 3$ and $0 < \beta < 2$ (Li [20]).

Class III contains vibrators with both fractional inertia force $m\dfrac{d^\alpha y(t)}{dt^\alpha}$ and fractional friction $c\dfrac{d^\beta y(t)}{dt^\beta}$, see, for example, Gomez-Aguilar [91, Eq. (10)], Tian et al. [92], Berman and Cederbaum [93], Coronel-Escamilla et al. [94, Eq. (12)], and Duan et al. [95, Eq. (2)]. Conventionally, in [35] and [91–95], fractional vibrators of (6.3) are for $1 < \alpha \leq 2$ and $0 < \beta \leq 1$. In our research, the developments for $1 < \alpha < 3$ and $0 < \beta < 2$ are considered (Li [20]).

6.2.4 Class IV Fractional Vibrators

We say that a fractional vibrator is in class IV if its motion equation is in the form of (6.4).

$$m\frac{d^{\alpha}x(t)}{dt^{\alpha}}+k\frac{d^{\lambda}x(t)}{dt^{\lambda}}=f(t),\quad 1<\alpha<3,\,0\leq\lambda<1. \qquad (6.4)$$

A class IV fractional vibrator IV is with fractional acceleration and fractional displacement. Reports about it are rarely seen but Li [20]. It may serve as a model to give a mathematical explanation of the standard expression of the Rayleigh damping assumption. Eq. (6.4) appears damping free in form, but it intrinsically contains a Rayleigh type damping (Li [20, 96–99]).

6.2.5 Class V Fractional Vibrators

We call a class V fractional vibrator if (6.5) is the expression of its motion equation in the form

$$m\frac{d^{2}x(t)}{dt^{2}}+k\frac{d^{\lambda}x(t)}{dt^{\lambda}}=f(t),\quad 0\leq\lambda<1. \qquad (6.5)$$

Mathematically, (6.5) is a special case for $\alpha=2$ in (6.4). It is with fractional restoration force. I take it as a class mainly from the point of view of vibration mechanics because the intrinsic damping in (6.5) is a variation of the Rayleigh damping (Li [20, 98]). Reports about it are seldom seen but Li [20, 96–99].

6.2.6 Class VI Fractional Vibrators

When the motion equation of a vibrator is given by (6.6) in the form

$$m\frac{d^{\alpha}y(t)}{dt^{\alpha}}+c\frac{d^{\beta}y(t)}{dt^{\beta}}+k\frac{d^{\lambda}y(t)}{dt^{\lambda}}=f(t),\quad 1<\alpha<3,\,0<\beta<2,\,0\leq\lambda<1, \qquad (6.6)$$

it is called a class VI fractional vibrator. It is with fractional inertia, fractional damping, and fractional restoration forces. It is rarely reported but Li [20, 96–99].

6.2.7 Class VII Fractional Vibrators

Introduce one more term for a class VII fractional vibrator. Eq. (6.7) is its expression of equation of motion in the form

$$m\frac{d^{2}x(t)}{dt^{2}}+c\frac{d^{\beta}x(t)}{dt^{\beta}}+k\frac{d^{\lambda}x(t)}{dt^{\lambda}}=f(t),\quad 0<\beta<2,\,0\leq\lambda<1. \qquad (6.7)$$

A class VII fractional vibration system is with fractional damping and fractional restoration forces. Reports regarding it are seldom seen but Li [96–99].

6.3 SUMMARY

We have described seven classes of fractional vibration systems with (6.1)–(6.7), respectively. Among them, reports about (6.4), (6.5), (6.6), and (6.7) are seldom seen, to the best of my knowledge.

REFERENCES

1. C. M. Harris, *Shock and Vibration Handbook*, 5th Ed., McGraw-Hill, New York, 2002.
2. O. M. Palley, Г. B. Bahizov, and E. Я. Voroneysk, *Handbook of Ship Structural Mechanics*, National Defense Industry Publishing House, Beijing, 2002. In Chinese. Translated from Russian by B. H. Xu, X. Xu, and M. Q. Xu.
3. J. J. Jensen, *Load and Global Response of Ships*, vol. 4, Elsevier, Academic Press, Oxford, 2001.
4. X. D. Jin and L. J. Xia, *Ship Hull Vibration*, The Press of Shanghai Jiaotong University, Shanghai, China, 2011. In Chinese.
5. H. A. Rothbart and Thomas H. Brown, Jr., *Mechanical Design Handbook*, 2nd Ed., Measurement, Analysis and Control of Dynamic Systems, McGraw–Hill, New York, 2006.
6. J. P. Den Hartog, *Mechanical Vibrations*, McGraw-Hill, New York, 1956.
7. A. A. Andronov, A. A. Victor, and C. Э. Hayijin, *Oscillation Theory*, Science Press, Beijing, China, 1974. In Chinese. Translated from Russian by W. B. Gao, R. W. Yang, and Z. Y. Xiao.
8. K. Nakagawa and M. Ringo, *Engineering Vibrations*, Shanghai Science and Technology Publishing House, Shanghai, China, 1981. In Chinese. Translated from Japanese by S. R. Xia.
9. C. Lalanne, *Mechanical Vibration and Shock*, Volume 1, Sinusoidal Vibration, 2nd Ed., John Wiley & Sons, Hoboken, 2013.
10. S. S. Ra, *Mechanical Vibrations*, 6th Ed., in SI Units, Pearson Education, Harlow, UK, 2018.
11. S. P. Timoshenko, *Mechanical Vibrations*, D. Van Nostrand Company, Inc., New York, 1955.
12. S. P. Timoshenko, *Vibration Problems in Engineering*, 2nd Ed., Fifth Printing, D. Van Nostrand Company, Inc., New York, 1955.
13. J. Falnes and A. Kurniawan, *Ocean Waves and Oscillating Systems: Linear Interactions Including Wave-Energy Extraction*, 2nd Ed., Cambridge University Press, Padstow Cornwall, 2020.
14. L. Banakh and M. Kempner, *Vibrations of Mechanical Systems with Regular Structure*, Foundations of Engineering Mechanics, Springer, Berlin, 2010.
15. E. N. Strømmen, *Structural Dynamics, Springer Series in Solid and Structural Mechanics*, vol. 2, Springer, Cham, 2014.

16. M. Mukhopadhyay, *Structural Dynamics*, Vibrations and Systems, Springer, Cham, 2021.
17. F. Cheli and G. Diana, *Advanced Dynamics of Mechanical Systems*, Springer, Cham, 2015.
18. R. Allemang and P. Avitabile, editors, *Handbook of Experimental Structural Dynamics*, 2 vols, Springer, New York, 2022.
19. I. Elishakoff, *Mechanical Vibration: Where Do We Stand?* vol. 488, International Centre for Mechanical Sciences, Springer, Berlin, 2007.
20. M. Li, *Fractional Vibrations with Applications to Euler-Bernoulli Beams*, CRC Press, Boca Raton, 2023.
21. A. C. J. Luo, *Dynamical Systems, Discontinuity, Stochasticity and Time-Delay*, Springer, New York, 2010.
22. B. Ross, *Fractional Calculus and Its Applications*, Lecture Notes in Mathematics, vol. 457, Springer, New York, 1975.
23. K. S. Miller and B. Ross, *An Introduction to the Fractional Calculus and Fractional Differential Equations*, John Wiley & Sons, New York, 1993.
24. B. Ross, Fractional calculus, *Mathematics Magazine*, vol. 50, no. 3, 1977, 115–122.
25. J. Klafter, S. C. Lim, and R. Metzler, *Fractional Dynamics: Recent Advances*, World Scientific, Singapore, 2012.
26. H. Weyl, Bemerkungen zum Begriff des Differentialquotienten gebrochener Ordnung, *Vierteljahrsschrift der Naturforschenden Gesellschaft in Zürich*, vol. 62, 1917, 296–302.
27. R. K. Raina and C. L. Koul, On Weyl fractional calculus, *Proceedings of the American Mathematical Society*, vol. 73, no. 2, 1979, 188–192.
28. T. M. Atanackovic, S. Pilipovic, B. Stankovic, and D. Zorica, *Fractional Calculus with Applications in Mechanics*, ISTE Ltd and John Wiley & Sons, London SW19 4EU/Hoboken, NJ 07030, 2014.
29. C. P. Li and F. H. Zeng, *Numerical Methods for Fractional Calculus*, Chapman and Hall/CRC, Boca Raton, 2015.
30. A. V. Pskhu and S. Sh. Rekhviashvili, Analysis of forced oscillations of a fractional oscillator, *Technical Physics Letters*, vol. 44, no. 12, 2018, 1218–1221.
31. Y. A. Rossikhin, Reflections on two parallel ways in progress of fractional calculus in mechanics of solids, *Applied Mechanics Reviews*, vol. 63, no. 1, 2010, 010701.
32. J. Čermák and T. Kisela, Stabilization and destabilization of fractional oscillators via a delayed feedback control, *Communications in Nonlinear Science and Numerical Simulation*, vol. 117, 2023, 106960.
33. J.-S. Duan, L. Jing, M. Li, The mixed boundary value problems and Chebyshev collocation method for Caputo-type fractional ordinary differential equations, *Fractal and Fractional*, vol. 6, no. 3, 2022, 148.
34. V. V. Uchaikin, Relaxation processes and fractional differential equations, *International Journal of Theoretical Physics*, vol. 42, no. 1, 2003, 121–134.
35. V. V. Uchaikin, *Fractional Derivatives for Physicists and Engineers*, vol. II, Springer, Berlin, 2013.
36. M. Li, Three classes of fractional oscillators, *Symmetry-Basel*, vol. 10, no. 2, 2018 (91 pages).

37. J.-S. Duan, The periodic solution of fractional oscillation equation with periodic input, *Advances in Mathematical Physics*, vol. 2013, 2013.
38. M. Zurigat, Solving fractional oscillators using Laplace homotopy analysis method, *Annals of the University of Craiova, Mathematics and Computer Science Series*, vol. 38, no. 4, 2011, 1–11.
39. T. Blaszczyk and M. Ciesielski, Fractional oscillator equation— Transformation into integral equation and numerical solution, *Applied Mathematics and Computation*, vol. 257, 2015, 428–435.
40. T. Blaszczyk, M. Ciesielski, M. Klimek, and J. Leszczynski, Numerical solution of fractional oscillator equation, *Applied Mathematics and Computation*, vol. 218, no. 6, 2011, 2480–2488.
41. T. Blaszczyk, A numerical solution of a fractional oscillator equation in a non-resisting medium with natural boundary conditions, *Romanian Reports in Physics*, vol. 67, no. 2, 2015, 350–358.
42. A. Al-Rabtah, V. S. Ertürk, and S. Momani, Solutions of a fractional oscillator by using differential transform method, *Computers & Mathematics with Applications*, vol. 59, no. 3, 2010, 1356–1362.
43. A. D. Drozdov, Fractional oscillator driven by a Gaussian noise, *Physica A*, vol. 376, 2007, 237–245.
44. A. A. Stanislavsky, Fractional oscillator, *Physical Review E*, vol. 70, no. 5, 2004, 051103 (6 pages).
45. A. N. N. Achar, J. W. Hanneken, and T. Clarke, Damping characteristics of a fractional oscillator, *Physica A*, vol. 339, no. 3–4, 2004, 311–319.
46. A. N. N. Achar, J. W. Hanneken, and T. Clarke, Response characteristics of a fractional oscillator, *Physica A*, vol. 309, no. 3–4, 2002, 275–288.
47. A. N. N. Achar, J. W. Hanneken, T. Enck, and T. Clarke, Dynamics of the fractional oscillator, *Physica A*, vol. 297, no. 3–4, 2001, 361–367.
48. A. Tofighi, The intrinsic damping of the fractional oscillator, *Physica A*, vol. 329, no. 1–2, 2003, 29–34.
49. Y. E. Ryabov and A. Puzenko, Damped oscillations in view of the fractional oscillator equation, *Physical Review B*, vol. 66, no. 18, 2002, 184201.
50. W. E. Ahmad and A. S. R. Elwakil, Fractional-order Wien-bridge oscillator, *Electronics Letters*, vol. 37, no. 18, 2001, 1110–1112.
51. J.-S. Duan, Z. Wang, Y.-L. Liu, and X. Qiu, Eigenvalue problems for fractional ordinary differential equations, *Chaos, Solitons & Fractals*, vol. 46, 2013, 46–53.
52. M. S. Tavazoei, Reduction of oscillations via fractional order pre-filtering, *Signal Processing*, vol. 107, 2015, 407–414.
53. T. Sandev and Z. Tomovski, The general time fractional wave equation for a vibrating string, *J. Physics A: Mathematical and Theoretical*, vol. 43, no. 5, 2010, 055204 (12pp).
54. H. Singh, H. M. Srivastava, and D. Kumar, A reliable numerical algorithm for the fractional vibration equation, *Chaos, Solitons & Fractals*, vol. 103, 2017, 131–138.
55. L.-F. Lin, C. Chen, S.-C. Zhong, and H.-Q. Wang, Stochastic resonance in a fractional oscillator with random mass and random frequency, *Journal of Statistical Physics*, vol. 160, no. 2, 2015, 497–511.

56. J.-S. Duan, A modified fractional derivative and its application to fractional vibration equation, *Applied Mathematics & Information Sciences*, vol. 10, no. 5, 2016, 1863–1869.

57. H. S. Alkhaldi, I. M. Abu-Alshaikh, and A. N. Al-Rabadi, Vibration control of fractionally-damped beam subjected to a moving vehicle and attached to fractionally-damped multi-absorbers, *Advances in Mathematical Physics*, vol. 2013, 2013.

58. H. Dai, Z. Zheng, and W. Wang, On generalized fractional vibration equation, *Chaos, Solitons & Fractals*, vol. 95, 2017, 48–51.

59. R. Ren, M. Luo, and K. Deng, Stochastic resonance in a fractional oscillator driven by multiplicative quadratic noise, *Journal of Statistical Mechanics: Theory and Experiment*, vol. 2017, 2017, 023210.

60. R. Ren, M. Luo, and K. Deng, Stochastic resonance in a fractional oscillator subjected to multiplicative trichotomous noise, *Nonlinear Dynamics*, vol. 90, no. 1, 2017, 379–390.

61. Y. Xu, Y. Li, D. Liu, W. Jia, and H. Huang, Responses of Duffing oscillator with fractional damping and random phase, *Nonlinear Dynamics*, vol. 74, no. 3, 2013, 745–753.

62. G. He, Y. Tian, and Y. Wang, Stochastic resonance in a fractional oscillator with random damping strength and random spring stiffness, *Journal of Statistical Mechanics: Theory and Experiment*, vol. 2013, 2013, P09026.

63. A. Y. T. Leung, Z. Guo, and H. X. Yang, Fractional derivative and time delay damper characteristics in Duffing-van der Pol oscillators, *Communications in Nonlinear Science and Numerical Simulation*, vol. 18, no. 10, 2013, 2900–2915.

64. L. C. Chen, Q. Q. Zhuang, and W. Q. Zhu, Response of SDOF nonlinear oscillators with lightly fractional derivative damping under real noise excitations, *The European Physical Journal Special Topics*, vol. 193, no. 1, 2011, 81–92.

65. J.-F. Deü and D. Matignon, Simulation of fractionally damped mechanical systems by means of a Newmark-diffusive scheme, *Computers & Mathematics with Applications*, vol. 59, no. 5, 2010, 1745–1753.

66. G. E. Drăgănescu, L. Bereteu, A. Ercuţa, and G. Luca, Anharmonic vibrations of a nano-sized oscillator with fractional damping, *Communications in Nonlinear Science and Numerical Simulation*, vol. 15, no. 4, 2010, 922–926.

67. Y. A. Rossikhin and M. V. Shitikova, New approach for the analysis of damped vibrations of fractional oscillators, *Shock and Vibration*, vol. 16, no. 4, 2009, 365–387.

68. F. Xie and X. Lin, Asymptotic solution of the van der Pol oscillator with small fractional damping, *Physica Scripta*, 2009 (T136) 2009, 014033 (4 pages).

69. J. Yuan, Y, Zhang, J, Liu, B, Shi, M, Gai, and S. Yang, Mechanical energy and equivalent differential equations of motion for single-degree-of-freedom fractional oscillators, *Journal of Sound and Vibration*, vol. 397, 2017, 192–203.

70. N. Nešić, D. Karličić, M. Cajić, J. Simonović, and S. Adhikari, Vibration suppression of a platform by a fractional type electromagnetic damper and inerter-based nonlinear energy sink, *Applied Mathematical Modelling*, vol. 137, January 2025, 115651.

71. L.-F. Lin, C. Chen, and H.-Q. Wang, Trichotomous noise induced stochastic resonance in a fractional oscillator with random damping and random frequency, *Journal of Statistical Mechanics: Theory and Experiment*, vol. 2016, 2016, 023201.

72. Y. Naranjani, Y. Sardahi, Y.-Q. Chen, and J.-Q. Sun, Multi-objective optimization of distributed-order fractional damping, *Communications in Nonlinear Science and Numerical Simulation*, vol. 24, no. 1–3, 2015, 159–168.

73. G. He, Y. Tian, and Y. Wang, Stochastic resonance in a fractional oscillator with random damping strength and random spring stiffness, *Journal of Statistical Mechanics: Theory and Experiment*, vol. 2013, 2013, P09026.

74. A. Di Matteo, P. D. Spanos, and A. Pirrotta, Approximate survival probability determination of hysteretic systems with fractional derivative elements, *Probabilistic Engineering Mechanics*, vol. 54, 2018, 138–146.

75. M. V. Shitikova, The fractional derivative expansion method in nonlinear dynamic analysis of structures, *Nonlinear Dynamics*, vol. 99, no. 1, 2020, 109–122.

76. B. Yıldız, S. Sınır, and B. G. Sınır, A general solution procedure for nonlinear single degree of freedom systems including fractional derivatives, *International Journal of Non-Linear Mechanics*, vol. 169, 2025, 104966.

77. R. Hu, D. Zhang, Z. Deng, and C. Xu, Stochastic analysis of a nonlinear energy harvester with fractional derivative damping, *Nonlinear Dynamics*, vol. 108, no. 3, 2022, 1973–1986.

78. Q. Cao, S.-L. James Hu, and H. Li, Frequency/Laplace domain methods for computing transient responses of fractional oscillators, *Nonlinear Dynamics*, vol. 108, no. 1, 2022, 1509–1523.

79. Q. Cao, S.-L. James Hu, and H. Li, Nonstationary response statistics of fractional oscillators to evolutionary stochastic excitation, *Communications in Nonlinear Science and Numerical Simulation*, vol. 103, 2021, 105962.

80. Q. B. Wang, H. Wu, and Y. J. Yang, The effect of fractional damping and time-delayed feedback on the stochastic resonance of asymmetric SD oscillator, *Nonlinear Dynamics*, vol. 107, 2022, 2099–2114.

81. B. Kaltenbacher and A. Schlintl, Fractional time stepping and adjoint based gradient computation in an inverse problem for a fractionally damped wave equation, *Journal of Computational Physics*, vol. 449, 2022, 110789.

82. P. D. Spanos and G. Malara, Nonlinear vibrations of beams and plates with fractional derivative elements subject to combined harmonic and random excitations, *Probabilistic Engineering Mechanics*, vol. 59, 2020, 103043.

83. G. Malara and P. D. Spanos, Nonlinear random vibrations of plates endowed with fractional derivative elements, *Probabilistic Engineering Mechanics*, vol. 54, 2018, 2–8.

84. Ž. Tomovski and T. Sandev, Effects of a fractional friction with power-law memory kernel on string vibrations, *Computers & Mathematics with Applications*, vol. 62, no. 3, 2011, 1554–1561.

85. V. M. Zelenev, S. I. Meshkov, and Y. A. Rossikhin, Damped vibrations of hereditary-elastic systems with weakly singular kernels, *Journal of Applied Mechanics and Technical Physics*, vol. 11, no. 2, 1970, 290–293.

86. Y. A. Rossikhin and M. V. Shitikova, Applications of fractional calculus to dynamic problems of linear and nonlinear hereditary mechanics of solids, *Applied Mechanics Reviews*, vol. 50, no. 1, 1997, 15–67.
87. Y. A. Rossikhin and M. V. Shitikova, Application of fractional calculus for dynamic problems of solid mechanics: Novel trends and recent results, *Applied Mechanics Reviews*, vol. 63, no. 1, 2010, 010801.
88. R. Bagley and P. J. Torvik, A generalized derivative model for an elastomer damper, *Shock and Vibration Bulletin*, vol. 49, no. 2, 1979, 135–143.
89. J.-S. Duan, Y.-J. Lan, M. Li, A comparative study of responses of fractional oscillator to sinusoidal excitation in the Weyl and Caputo senses, *Fractal and Fractional*, vol. 6, no. 12, 2022, 692.
90. G. P. Tsinker, *Marine Structures Engineering: Specialized Applications*, Springer, Berlin, 1995.
91. J. F. Gomez-Aguilar, J. J. Rosales-Garcia, J. J. Bernal-Alvarado, T. Cordova-Fraga, and R. Guzman-Cabrera, Fractional mechanical oscillators, *Revista Mexicana de Fisica*, vol. 58, no. 4, 2012, 348–352.
92. Y. Tian, L.-F. Zhong, G.-T. He, T. Yu, M.-K. Luo, and H. E. Stanley, The resonant behavior in the oscillator with double fractional-order damping under the action of nonlinear multiplicative noise, *Physica A*, vol. 490, 2018, 845–856.
93. M. Berman and L. S. Cederbaum, Fractional driven-damped oscillator and its general closed form exact solution, *Physica A*, vol. 505, 2018, 744–762.
94. A. Coronel-Escamilla, J. F. Gómez-Aguilar, D. Baleanu, T. Córdova-Fraga, R. F. Escobar-Jiménez, V. H. Olivares-Peregrino, and M. M. Al Qurashi, Bateman-Feshbach Tikochinsky and Caldirola-Kanai oscillators with new fractional differentiation, *Entropy*, vol. 19, no. 2, 2017 (55 pages).
95. J.-S. Duan, M. Li, Y. Wang, and Y.-L. An, Approximate solution of fractional differential equation by quadratic splines, *Fractal and Fractional*, vol. 6, no. 7, 2022, 369.
96. M. Li, Stationary responses of seven classes of fractional vibrations driven by sinusoidal force, *Fractal and Fractional*, vol. 8, no. 8, 2024, 479.
97. M. Li, Dealing with stationary sinusoidal responses of seven types of multi-fractional vibrators using multi-fractional phasor, *Symmetry*, vol. 16, no. 9, 2024, 1197.
98. M. Li, Analytic theory of seven classes of fractional vibrations based on elementary functions: A tutorial review, *Symmetry*, vol. 16, no. 9, 2024, 1202.
99. M. Li, PSD and cross PSD of responses of seven classes of fractional vibrations driven by fGn, fBm, fractional OU process, and von Kármán process, *Symmetry*, vol. 16, no. 5, 2024, 635.

Seven Classes of Fractional Vibrations

THIS CHAPTER IS A survey about the closed-form expressions of equivalent motion equations, equivalent masses, equivalent dampings, equivalent stiffnesses, equivalent damping ratios, equivalent damping free natural angular frequencies, equivalent damped natural angular frequencies, equivalent frequency ratios, responses (free, impulse, and step), frequency transfer functions, equivalent logarithmic decrements, and equivalent Q factors with respect to the fractional vibration systems from class I to class VI using elementary functions.

7.1 BACKGROUND

The literature of fractional vibrations is rich, see, for example, Uchaikin [1], Shitikova [2, 3], Spanos and Malara [4–6], Duan [7, 8], Duan et al. [9–13], Zurigat [14], Blaszczyk and Ciesielski [15], Blaszczyk et al. [16], Blaszczyk [17], Al-Rabtah [18], Drozdov [19], Stanislavsky [20], Achar et al. [21–23], Tofighi [24], Ryabov and Puzenko [25], Tavazoei [26], Sandev and Tomovski [27], Singh et al. [28], Lin et al. [29, 30], Alkhaldi et al. [31], Dai et al. [32], Ren et al. [33, 34], Xu et al. [35], He et al. [36], Leung et al. [37], Chen et al. [38], Deü and Matignon [39], Drăgănescu et al. [40], Rossikhin and Shitikova [41], Xie and Lin [42], Yuan et al. [43], Naranjani et al. [44], Matteo et al. [45], Tomovski and Sandev [46], Gomez-Aguilar et al. [47], Tian et al. [48], Berman and Cederbaum [49], and Coronel-Escamilla et al. [50], just mentioning a few. Nevertheless, the analytic expressions

DOI: 10.1201/9781003657897-7

using elementary functions with respect to responses are rarely reported, letting along the analytic expressions of responses to fractional random vibrations.

Note that the analytically closed-form expressions of vibration responses of six classes of fractional vibration systems were recently introduced by Li [51–53]. In addition, the analytically closed-form expressions of vibration responses of seven classes of Euler-Bernoulli beams were brought forward by Li [51]. One of the themes of this book is to introduce the analytically closed-form expressions of fractional random vibrations of seven classes of fractional vibration systems addressed in this chapter. The highlights in this chapter are in two aspects. One is a survey of six classed fractional vibrations discussed in Li [51–53]. The other is to newly introduce the class VII fractional vibration systems. The details about the proofs of those results for classes I–VI fractional vibrators refer to Li [51–53]. The proofs of the results regarding class VII fractional vibrators are given in this chapter and also Li [54].

7.2 RESULTS FOR CLASS I FRACTIONAL VIBRATION SYSTEMS

7.2.1 Equivalent Motion Equation of a Class I Fractional Vibrator

The motion equation of a class I fractional vibrator is expressed by

$$B_1(t) \triangleq m \frac{d^\alpha x_1(t)}{dt^\alpha} + kx_1(t) = 0. \tag{7.1}$$

In (7.1), $1 < \alpha < 3$. Eq. (7.2) is the equivalent equation of (7.1). It is given by

$$A_1(t) \triangleq -m\omega^{\alpha-2} \cos\frac{\alpha\pi}{2} \frac{d^2 x_1(t)}{dt^2} + m\omega^{\alpha-1} \sin\frac{\alpha\pi}{2} \frac{dx_1(t)}{dt} + kx_1(t) = 0. \tag{7.2}$$

In fact, $F[A_1(t) - B_1(t)] = 0$, where F is the operator of Fourier transform.

7.2.2 Equivalent Vibration Parameters of Class I Fractional Vibrators

7.2.2.1 Equivalent Mass of Class I Fractional Vibrators

Denote by m_{eq1} the equivalent mass for a fractional vibrator of class I. From (7.2), m_{eq1} is expressed by

$$m_{\mathrm{eq1}} = -m\omega^{\alpha-2} \cos\frac{\alpha\pi}{2}. \tag{7.3}$$

Eq. (7.3) implies that $m_{eq1} = m_{eq1}(\omega, \alpha, m)$. It exhibits that $m_{eq1}(\omega, \alpha, m) \geq 0$ for $1 < \alpha < 3$. It reduces to the primary mass m when $\alpha = 2$. It has two asymptotic properties described by

$$\lim_{\omega \to \infty} m_{eq1}(\omega, \alpha, m) = 0, \quad 1 < \alpha < 2, \tag{7.4}$$

$$\lim_{\omega \to \infty} m_{eq1}(\omega, \alpha, m) = \infty, \quad 2 < \alpha < 3. \tag{7.5}$$

In vibrations, $\omega \to \infty$ implies high-order vibrations modes and coordinates of high-order vibrations modes approach zero in general. Thus, $\omega \to \infty$ may usually be trivial in vibrations. However, a class I fractional vibrator has properties substantially different from an ordinary vibrator in mass. When $1 < \alpha < 2$, $m_{eq1}(\omega, \alpha, m)$ disappears for $\omega \to \infty$ as can be seen from (7.4). On the other hand, (7.5) says that m_{eq1} becomes ∞ for $\omega \to \infty$ if $2 < \alpha < 3$.

In addition, there are two asymptotic behaviours of m_{eq1} for $\omega \to 0$. They are described by

$$\lim_{\omega \to 0} m_{eq1}(\omega, \alpha, m) = \infty, \quad 1 < \alpha < 2, \tag{7.6}$$

$$\lim_{\omega \to 0} m_{eq1}(\omega, \alpha, m) = 0, \quad 2 < \alpha < 3. \tag{7.7}$$

Eqs. (7.6) and (7.7) imply that $0 < m_{eq1} < \infty$. Conventional vibrations do not have such a phenomenon of mass.

7.2.2.2 Equivalent Damping of Class I Fractional Vibrators

Let c_{eq1} be the equivalent damping for a class I fractional vibrator. From (7.2), we have

$$c_{eq1} = m\omega^{\alpha-1} \sin \frac{\alpha \pi}{2}. \tag{7.8}$$

From (7.8), we see that when $\alpha = 2$, c_{eq1} vanishes. When $1 < \alpha < 2$, c_{eq1} is positive definite. That is,

$$c_{eq1} \geq 0. \tag{7.9}$$

In passing, we say that c_{eq1} is proportional to the primary mass m. That is just a mathematical explanation of a variation of the Rayleigh damping assumption.

The quantity c_{eq1} may be negative. In fact, for $2 < \alpha < 3$ and $\omega \neq 0$, we have

$$c_{eq1} < 0. \tag{7.10}$$

Write c_{eq1} by $c_{eq1}(\omega, \alpha, m)$. We have its asymptotic properties given by

$$\lim_{\omega \to 0} c_{eq1}(\omega, \alpha, m) = 0, \quad 1 < \alpha < 2, \tag{7.11}$$

$$\lim_{\omega \to \infty} c_{eq1}(\omega, \alpha, m) = \infty, \tag{7.12}$$

$$\lim_{\omega \to 0} c_{eq1}(\omega, \alpha, m) = 0, \quad 2 < \alpha < 3, \tag{7.13}$$

$$\lim_{\omega \to \infty} c_{eq1}(\omega, \alpha, m) = -\infty. \tag{7.14}$$

Eqs. (7.11)–(7.14) exhibit that $-\infty < c_{eq1} < \infty$.

7.2.2.3 Equivalent Damping Ratio of Class I Fractional Vibrators

Denote by ζ_{eq1} the equivalent damping ratio for a class I fractional vibrator. Define it by (7.15) in the form

$$\zeta_{eq1} = \frac{c_{eq1}}{2\sqrt{m_{eq1}k}}. \tag{7.15}$$

According to (7.3) and (7.8), we have

$$\zeta_{eq1} = \zeta_{eq1}(\omega, \alpha) = \frac{\omega^{\frac{\alpha}{2}} \sin \frac{\alpha\pi}{2}}{2\omega_n \sqrt{-\cos \frac{\alpha\pi}{2}}}. \tag{7.16}$$

In (7.16), $\omega_n = \sqrt{\dfrac{k}{m}}$ is the conventional natural angular frequency with damping free.

Note that the $\zeta_{eq1} = 0$ when $\alpha = 2$. The properties of ζ_{eq1} in terms of α are explained by (7.17) and (7.18).

- When $1 < \alpha < 2$,

$$\varsigma_{eq1} \geq 0. \tag{7.17}$$

- If $2 < \alpha < 3$ and $\omega \neq 0$,

$$\varsigma_{eq1} < 0. \tag{7.18}$$

A class I fractional vibrator appears damping free in form. However, fractional inertia force $m \dfrac{d^{\alpha} x(t)}{dt^{\alpha}}$ produces c_{eq1} and accordingly ς_{eq1}. They may be positive, negative or zero, relying on α and ω.

7.2.2.4 Equivalent Damping Free Natural Frequency of Class I Fractional Vibrators

Denote by ω_{eqn1} the equivalent damping-free natural angular frequency for a class I fractional vibrator. Define it by

$$\omega_{eqn1} = \sqrt{\frac{k}{m_{eq1}}}. \tag{7.19}$$

Substituting ς_{eq1} in (7.3) into (7.19) produces

$$\omega_{eqn1} = \frac{\omega_n}{\sqrt{-\omega^{\alpha-2} \cos \dfrac{\alpha \pi}{2}}}. \tag{7.20}$$

From (7.20), we see that, in general, ω_{eqn1} is not a constant but a function of ω and α. It reduces to $\omega_n = \text{const}$ if and only if $\alpha = 2$.

7.2.2.5 Equivalent Damped Natural Frequency of Class I Fractional Vibrators

Let ω_{eqd1} be the equivalent damped natural angular frequency for a class I fractional vibrator. Since $\varsigma_{eq1} > 1$ is trivial in vibrations, we restrict $|\varsigma_{eq1}| \leq 1$ in what follows. Define

$$\omega_{eqd1} = \omega_{eqn1} \sqrt{1 - \varsigma_{eq1}^2}. \tag{7.21}$$

Substituting ζ_{eq1} in (7.16) and ω_{eqn1} in (7.20) into (7.21) produces

$$\omega_{eqd1} = \frac{\omega_n}{\sqrt{-\omega^{\alpha-2}\cos\dfrac{\alpha\pi}{2}}}\sqrt{1 - \frac{\omega^\alpha \sin^2 \dfrac{\alpha\pi}{2}}{4\omega_n^2 \left|\cos \dfrac{\alpha\pi}{2}\right|}}. \tag{7.22}$$

7.2.2.6 Equivalent Frequency Ratio of Class I Fractional Vibrators
Let γ_{eq1} be the equivalent frequency ratio of a class I fractional vibrator. It is defined by

$$\gamma_{eq1} = \frac{\omega}{\omega_{eqn1}}. \tag{7.23}$$

Substituting ω_{eqn1} in (7.20) into (7.23) yields

$$\gamma_{eq1} = \gamma_{eq1}(\omega,\alpha) = \frac{\omega\sqrt{-\omega^{\alpha-2}\cos\dfrac{\alpha\pi}{2}}}{\omega_n}. \tag{7.24}$$

From (7.24), we see that γ_{eq1} is not dimensionless in general. It reduces to being dimensionless if and only if $\alpha = 2$. In fact, $\gamma_{eq1}(\omega,2) = \gamma = \dfrac{\omega}{\omega_n}$.

7.2.3 Responses of Class I Fractional Vibrators
7.2.3.1 Free Response of Class I Fractional Vibrators
Denote by $x_1(t)$ the free response of a class I fractional vibrator. It is the solution

$$\begin{cases} m\dfrac{d^\alpha x_1(t)}{dt^\alpha} + kx_1(t) = 0, \\ x_1(0) = x_{10}, x_1'(0) = v_{10}. \end{cases} \tag{7.25}$$

Since $F[A_1(t) - B_1(t)] = 0$, the solution to (7.25) equals to the solution to the one to

$$\begin{cases} m_{eq1}\dfrac{d^2 x_1(t)}{dt^2} + c_{eq1}\dfrac{dx_1(t)}{dt} + kx_1(t) = 0, \\ x_1(0) = x_{10}, x_1'(0) = v_{10}. \end{cases} \tag{7.26}$$

From (7.26), we write the solution to (7.26) by

$$x_1(t) = e^{-\zeta_{eq1}\omega_{eqn1}t}\left(x_{10}\cos\omega_{eqd1}t + \frac{v_{10} + \zeta_{eq1}\omega_{eqn1}x_{10}}{\omega_{eqd1}}\sin\omega_{eqd1}t\right), \quad t \geq 0. \quad (7.27)$$

Substituting ζ_{eq1}, ω_{eqn1}, and ω_{eqd1} into (7.27) yields

$$x_1(t) = e^{-\frac{\omega^{\frac{\alpha}{2}}\sin\frac{\alpha\pi}{2}}{2\left|\cos\frac{\alpha\pi}{2}\right|\sqrt{\left|\omega^{\alpha-2}\cos\frac{\alpha\pi}{2}\right|}}t}\begin{bmatrix}\left(x_{10}\cos\dfrac{\omega_n}{\sqrt{\left|\omega^{\alpha-2}\right|\cos\frac{\alpha\pi}{2}}}\sqrt{1 - \dfrac{\omega^{\alpha}\sin^2\frac{\alpha\pi}{2}}{4\omega_n^2\left|\cos\frac{\alpha\pi}{2}\right|}}t\right) \\ + \dfrac{v_{10} + \dfrac{\omega^{\frac{\alpha}{2}}\sin\frac{\alpha\pi}{2}}{2\sqrt{\left|\cos\frac{\alpha\pi}{2}\right|}}\dfrac{x_{10}}{\sqrt{\left|\omega^{\alpha-2}\right|\cos\frac{\alpha\pi}{2}}}}{\dfrac{\omega_n}{\sqrt{\left|\omega^{\alpha-2}\right|\cos\frac{\alpha\pi}{2}}}\sqrt{1 - \dfrac{\omega^{\alpha}\sin^2\frac{\alpha\pi}{2}}{4\omega_n^2\left|\cos\frac{\alpha\pi}{2}\right|}}} \\ \cdot\sin\left(\dfrac{\omega_n}{\sqrt{\left|\omega^{\alpha-2}\right|\cos\frac{\alpha\pi}{2}}}\sqrt{1 - \dfrac{\omega^{\alpha}\sin^2\frac{\alpha\pi}{2}}{4\omega_n^2\left|\cos\frac{\alpha\pi}{2}\right|}}t\right)\end{bmatrix}. \quad (7.28)$$

7.2.3.2 Impulse Response of Class I Fractional Vibrators

Let $h_1(t)$ be the impulse response of a class I fractional vibrator. It is the solution to the fractional differential equation given by

$$\begin{cases} m\dfrac{d^{\alpha}h_1(t)}{dt^{\alpha}} + kh_1(t) = \delta(t), \\ h_1(0) = 0, h_1'(0) = 0. \end{cases} \quad (7.29)$$

Due to $F[A_1(t) - B_1(t)] = 0$, the solution to (7.29) is equivalent to the one to the differential equation in the form

$$\begin{cases} m_{eq1}\dfrac{d^2h_1(t)}{dt^2} + c_{eq1}\dfrac{dh_1(t)}{dt} + kh_1(t) = \delta(t), \\ h_1(0) = 0, h_1'(0) = 0. \end{cases} \quad (7.30)$$

From (7.30), we have

$$h_1(t) = \frac{e^{-\varsigma_{eq1}\omega_{eqn1}t}}{m_{eq1}\omega_{eqd1}}\sin\omega_{eqd1}t, \quad t \geq 0. \qquad (7.31)$$

Substituting m_{eq1}, ς_{eq1}, ω_{eqn1}, and ω_{eqd1} into (7.31) results in

$$h_1(t) = \frac{e^{-\frac{\omega\sin\frac{\alpha\pi}{2}}{2\left|\cos\frac{\alpha\pi}{2}\right|}t}\sin\left(\frac{\omega_n}{\sqrt{\omega^{\alpha-2}\left|\cos\frac{\alpha\pi}{2}\right|}}\sqrt{1-\frac{\omega^{2\alpha}\sin^2\frac{\alpha\pi}{2}}{4\omega_n^2\left|\cos\frac{\alpha\pi}{2}\right|}}t\right)}{m\omega_n\sqrt{\omega^{\alpha-2}\left|\cos\frac{\alpha\pi}{2}\right|}\sqrt{1-\frac{\omega^{2\alpha}\sin^2\frac{\alpha\pi}{2}}{4\omega_n^2\left|\cos\frac{\alpha\pi}{2}\right|}}}, \quad t \geq 0. \qquad (7.32)$$

7.2.3.3 Step Response of Class I Fractional Vibrators

Denote by $g_1(t)$ the step response of a class I fractional vibrator. It is the solution to

$$\begin{cases} m\dfrac{d^\alpha g_1(t)}{dt^\alpha} + kg_1(t) = u(t), \\ g_1(0) = 0, g_1'(0) = 0. \end{cases} \qquad (7.33)$$

The solution to (7.33) equals to the one to

$$\begin{cases} m_{eq1}\dfrac{d^2 g_1(t)}{dt^2} + c_{eq1}\dfrac{dg_1(t)}{dt} + kg_1(t) = u(t), \\ g_1(0) = 0, g_1'(0) = 0. \end{cases} \qquad (7.34)$$

From (7.34), we have

$$g_1(t) = \int_0^t h_1(\tau)d\tau = \frac{1}{k}\left[1 - \frac{e^{-\varsigma_{eq1}\omega_{eqn1}t}\omega_{eqn1}}{\omega_{eqd1}}\cos\left(\omega_{eqd1}t - \phi_1\right)\right], \quad t \geq 0. \qquad (7.35)$$

Eq. (7.36) shows the expression of ϕ_1 in (7.35) in the form

$$\phi_1 = \tan^{-1}\frac{\varsigma_{eq1}}{\sqrt{1-\varsigma_{eq1}^2}}. \qquad (7.36)$$

Substituting ζ_{eq1}, ω_{eqn1}, and ω_{eqd1} into (7.35) produces

$$g_1(t) = \frac{1}{k}\left[1 - \frac{1}{\sqrt{1-\zeta_{eq1}^2}}e^{-\frac{\omega\sin\frac{\alpha\pi}{2}}{2\left|\cos\frac{\alpha\pi}{2}\right|}t}\cos\left(\frac{\omega_n}{\sqrt{-\omega^{\alpha-2}\cos\frac{\alpha\pi}{2}}}\sqrt{1-\frac{\omega^\alpha\sin^2\frac{\alpha\pi}{2}}{4\omega_n^2\left|\cos\frac{\alpha\pi}{2}\right|}}t - \phi_1\right)\right]. \qquad (7.37)$$

7.2.4 Frequency Transfer Function of Class I Fractional Vibrators

Let $H_1(\omega)$ be the frequency transfer function of a class I fractional vibrator. It is given by

$$H_1(\omega) = \frac{1}{k\left(1 - \frac{\omega^\alpha}{\omega_n^2}\left|\cos\frac{\alpha\pi}{2}\right| + i\frac{\omega^\alpha}{\omega_n^2}\sin\frac{\alpha\pi}{2}\right)}. \qquad (7.38)$$

Rewrite (7.38) by

$$H_1(\omega) = \frac{1}{k\left(1 - \gamma_{eq1}^2 + i2\varsigma_{eq1}\gamma_{eq1}\right)}. \qquad (7.39)$$

In the polar system, $H_1(\omega) = |H_1(\omega)|\exp[-\phi_1(\omega)]$. The amplitude of $H_1(\omega)$ is given by

$$|H_1(\omega)| = \frac{1/k}{\sqrt{\left(1 - \frac{\omega^\alpha}{\omega_n^2}\left|\cos\frac{\alpha\pi}{2}\right|\right)^2 + \left(\frac{\omega^\alpha}{\omega_n^2}\sin\frac{\alpha\pi}{2}\right)^2}}. \qquad (7.40)$$

Eq. (7.40) can be rewritten by

$$|H_1(\omega)| = \frac{1/k}{\sqrt{\left(1 - \gamma_{eq1}^2\right)^2 + \left(2\varsigma_{eq1}\gamma_{eq1}\right)^2}}. \qquad (7.41)$$

Eq. (7.42) shows $\phi_1(\omega)$ in the form

$$\varphi_1(\omega) = \tan^{-1}\frac{\omega^\alpha\sin\frac{\alpha\pi}{2}}{\omega_n^2 - \omega^\alpha\left|\cos\frac{\alpha\pi}{2}\right|}. \qquad (7.42)$$

For the purpose of computation of $\phi_1(\omega)$ with digital computers, we use

$$\varphi_1(\omega) = \cos^{-1} \frac{1-\gamma_{eq1}^2}{\sqrt{\left(1-\gamma_{eq1}^2\right)^2 + \left(2\varsigma_{eq1}\gamma_{eq1}\right)^2}}. \tag{7.43}$$

7.2.5 Equivalent Logarithmic Decrement and Q Factor of Class I Fractional Vibrators

7.2.5.1 Equivalent Logarithmic Decrement of Class I Fractional Vibrators

Let t_i and t_{i+1} be two successive time points of the free response $x_1(t)$, where $x_1(t)$ reaches its successive peak values of $x_1(t_i)$ and $x_1(t_{i+1})$, respectively. Let Δ_{eq1} be the equivalent logarithmic decrement of $x_1(t)$. Then,

$$\Delta_{eq1} = \Delta_{eq1}(\omega, \alpha) = \frac{\pi}{\sqrt{1 - \left(\frac{\omega^{\frac{\alpha}{2}} \sin\frac{\alpha\pi}{2}}{2\omega_n\sqrt{-\cos\frac{\alpha\pi}{2}}}\right)^2}} \frac{\omega^{\frac{\alpha}{2}} \sin\frac{\alpha\pi}{2}}{\omega_n\sqrt{-\cos\frac{\alpha\pi}{2}}}. \tag{7.44}$$

Note that the conventional logarithmic decrement is dimensionless. However, Δ_{eq1} is not. In addition, conventionally, logarithmic decrement is a constant. Nonetheless, $\Delta_{eq1}(\omega, \alpha)$ is a function of ω and α. When $\alpha = 2$, $\Delta_{eq1}(\omega, 2) = 0$.

7.2.5.2 Equivalent Q Factor of Class I Fractional Vibrators

Let Q_{eq1} be the equivalent Q factor of a class I fractional vibrator. Then,

$$Q_{eq1} = Q_{eq1}(\omega, \alpha) = \frac{\omega_n\sqrt{-\cos\frac{\alpha\pi}{2}}}{\omega^{\frac{\alpha}{2}} \sin\frac{\alpha\pi}{2}}. \tag{7.45}$$

Note that Q_{eq1} is usually not a constant as that of the conventional one. Instead, it is a function with respect to ω and α. Besides, the conventional Q is dimensionless but Q_{eq1} is not. If $\alpha = 2$, Q_{eq1} is dimensionless and $Q_{eq1}(\omega, 2) = \infty$.

7.3 RESULTS FOR CLASS II FRACTIONAL VIBRATION SYSTEMS

7.3.1 Equivalent Motion Equation of Class II Fractional Vibrators

The motion equation of a class II fractional vibrator is given by

$$B_2(t) \triangleq m\frac{d^2x_2(t)}{dt^2} + c\frac{d^\beta x_2(t)}{dt^\beta} + kx_2(t) = 0. \qquad (7.46)$$

In (7.46), $0 < \beta < 2$. Eq. (7.47) is the equivalent equation of (7.46). It is given by

$$A_2(t) \triangleq \left(m - c\omega^{\beta-2}\cos\frac{\beta\pi}{2}\right)\frac{d^2x_2(t)}{dt^2} + \left(c\omega^{\beta-1}\sin\frac{\beta\pi}{2}\right)\frac{dx_2(t)}{dt} + kx_2(t) = 0. \qquad (7.47)$$

In fact, $F[A_2(t) - B_2(t)] = 0$.

7.3.2 Equivalent Vibration Parameters of Class II Fractional Vibrators

7.3.2.1 Equivalent Mass of Class II Fractional Vibrators

Denote by m_{eq2} the equivalent mass for a class II fractional vibrator. From (7.47), therefore, we see that m_{eq2} is expressed by

$$m_{eq2} = m_{eq2}(\omega, \beta) = m - c\omega^{\beta-2}\cos\frac{\beta\pi}{2}. \qquad (7.48)$$

Eq. (7.48) exhibits that m_{eq2} relates to ω and β. It consists of two parts. One is the primary mass m and the other $-c\omega^{\beta-2}\cos\frac{\beta\pi}{2}$. The latter is contributed by fractional friction force $c\frac{d^\beta y(t)}{dt^\beta}$. In other words, fractional friction force may produce a certain added mass. Taking into account $\frac{c}{m} = 2\varsigma\omega_n$, we rewrite (7.48) by

$$m_{eq2} = m\left(1 - 2\varsigma\omega_n\omega^{\beta-2}\cos\frac{\beta\pi}{2}\right). \qquad (7.49)$$

From (7.48) or (7.49), we infer the interesting asymptotic properties regarding m_{eq2} by

$$\lim_{\omega \to \infty} m_{eq2}(\omega,\beta,m) = m, \tag{7.50}$$

$$\lim_{\omega \to 0} m_{eq2}(\omega,\beta,m) = -\infty, \quad 0 < \beta < 1. \tag{7.51}$$

The condition for m_{eq2} being negative is expressed by

$$2\varsigma\omega_n\omega^{\beta-2}\cos\frac{\beta\pi}{2} > 1. \tag{7.52}$$

Eqs. (7.48)–(7.51) exhibit that the range of m_{eq2} is $(-\infty, m)$. That is a particularly interesting phenomenon.

7.3.2.2 Equivalent Damping of Class II Fractional Vibrators

Let c_{eq2} be the equivalent damping for a class II fractional vibrator. As can be seen from (7.47), we write c_{eq2} by

$$c_{eq2} = c_{eq2}(\omega,\beta) = c\omega^{\beta-1}\sin\frac{\beta\pi}{2}. \tag{7.53}$$

From (7.53), we see that c_{eq2} reduces to the primary damping c when $\beta = 1$. Eq. (7.53) exhibits that it is positive definite for $0 < \beta < 2$. Precisely,

$$c_{eq2} \geq 0. \tag{7.54}$$

Asymptotic properties of c_{eq2} are expressed by

$$\lim_{\omega \to 0} c_{eq2}(\omega,\beta) = \infty, \quad 0 < \beta < 1, \tag{7.55}$$

$$\lim_{\omega \to \infty} c_{eq2}(\omega,\beta) = 0, \quad 0 < \beta < 1, \tag{7.56}$$

$$\lim_{\omega \to 0} c_{eq2}(\omega,\beta) = 0, \quad 1 < \beta < 2, \tag{7.57}$$

$$\lim_{\omega \to \infty} c_{eq2}(\omega,\beta) = \infty, \quad 1 < \beta < 2. \tag{7.58}$$

Eqs. (7.55)–(7.58) imply that the range of c_{eq2} is $(0, \infty)$.

7.3.2.3 Equivalent Damping Ratio of Class II Fractional Vibrators

Let ζ_{eq2} be the equivalent damping ratio for a class II fractional vibrator. Define it by

$$\zeta_{eq2} = \frac{c_{eq2}}{2\sqrt{m_{eq2}k}}. \tag{7.59}$$

Substituting m_{eq2} in (7.48) and c_{eq2} in (7.53) into (7.59) results in

$$\zeta_{eq2} = \zeta_{eq2}(\omega, \beta) = \frac{\varsigma \omega^{\beta-1} \sin\dfrac{\beta\pi}{2}}{\sqrt{1 - \dfrac{c}{m}\omega^{\beta-2}\cos\dfrac{\beta\pi}{2}}}. \tag{7.60}$$

In (7.60), $\varsigma = \dfrac{c}{2\sqrt{mk}}$. The quantity ζ_{eq2} is·not dimensionless in general. It is dimensionless if and only if $\beta = 1$. It is positive definite. That is,

$$\zeta_{eq2} \geq 0. \tag{7.61}$$

7.3.2.4 Equivalent Damping-Free Natural Frequency of Class II Fractional Vibrators

Let ω_{eqn2} be the equivalent damping-free natural angular frequency for a class II fractional vibrator. It is defined by

$$\omega_{eqn2} = \sqrt{\frac{k}{m_{eq2}}}.$$

Substituting m_{eq2} into the previous equation produces

$$\omega_{eqn2} = \frac{\omega_n}{\sqrt{1 - \dfrac{c}{m}\omega^{\beta-2}\cos\dfrac{\beta\pi}{2}}}. \tag{7.62}$$

The unit of ω_{eqn2} is generally not rad/s. Its unit is rad/s if and only $\beta = 1$ or $\omega \to \infty$.

7.3.2.5 Equivalent Damped Natural Frequency of Class II Fractional Vibrators

Without losing the generality, small damping $|\zeta_{eq2}| \leq 1$ is assumed from a view of vibration engineering. Denote the equivalent damped natural angular frequency for a class II fractional vibrator as ω_{eqd2}. Define it by

$$\omega_{eqd2} = \omega_{eqn2} \sqrt{1 - \zeta_{eq2}^2}. \tag{7.63}$$

Substituting ω_{eqn2} and ζ_{eq2} into (7.63) yields

$$\omega_{eqd2} = \frac{\omega_n}{\sqrt{1 - \dfrac{c}{m}\omega^{\beta-2}\cos\dfrac{\beta\pi}{2}}} \sqrt{1 - \frac{\zeta^2 \omega^{2(\beta-1)}\sin^2\dfrac{\beta\pi}{2}}{1 - \dfrac{c}{m}\omega^{\beta-2}\cos\dfrac{\beta\pi}{2}}}. \tag{7.64}$$

The unit of ω_{eqd2} is not rad/s unless $\beta = 1$ or $\omega \to \infty$.

7.3.2.6 Equivalent Frequency Ratio of Class II Fractional Vibrators

Let γ_{eq2} be the equivalent frequency ratio for a class II fractional vibrator. It is defined by

$$\gamma_{eq2} = \frac{\omega}{\omega_{eqn2}}. \tag{7.65}$$

Substituting ω_{eqn2} into (7.65) produces

$$\gamma_{eq2} = \gamma_{eq2}(\omega, \beta) = \gamma\sqrt{1 - \frac{c}{m}\omega^{\beta-2}\cos\frac{\beta\pi}{2}}. \tag{7.66}$$

In (7.66), $\gamma = \dfrac{\omega}{\omega_n}$. The unit of γ_{eq2} is not dimensionless. It reduces to being dimensionless when $\beta = 1$ or $\omega \to \infty$.

7.3.3 Responses of Class II Fractional Vibration Systems

7.3.3.1 Free Response of Class II Fractional Vibrators

Let $x_2(t)$ be the free response of a class II fractional vibrator. It is the solution to the fractional differential equation given by

$$\begin{cases} m\dfrac{d^2 x_2(t)}{dt^2} + c\dfrac{d^\beta x_2(t)}{dt^\beta} + kx_2(t) = 0, \\ x_2(0) = x_{20}, x_2'(0) = v_{20}. \end{cases} \tag{7.67}$$

Because of $F[A_2(t) - B_2(t)] = 0$, the solution to (7.67) is equivalent to the one to

$$\begin{cases} \dfrac{d^2 x_2(t)}{dt^2} + 2\varsigma_{eq2}\omega_{eqn2}\dfrac{dx_2(t)}{dt} + \omega_{eqn2}^2 x_2(t) = 0, \\ x_2(0) = x_{20}, x_2'(0) = v_{20}. \end{cases} \tag{7.68}$$

Eq. (7.69) is the solution to (7.68). It is in the form

$$x_2(t) = e^{-\varsigma_{eq2}\omega_{eqn2}t}\left[x_{20}\cos\omega_{eqd2}t + \dfrac{v_{20} + \varsigma_{eq2}\omega_{eqn2}x_{20}}{\omega_{eqd2}}\sin\omega_{eqd2}t \right], \quad t \geq 0. \tag{7.69}$$

Substituting $\varsigma_{eq2}(\omega, \beta)$ and $\omega_{eqd2}(\omega, \beta)$ into (7.69) yields

$$x_2(t) = e^{\frac{\varsigma_n\omega^{\beta-1}\sin\frac{\beta\pi}{2}}{1-\frac{c}{m}\omega^{\beta-2}\cos\frac{\beta\pi}{2}}t}\left[\begin{array}{l} x_{20}\cos\left(\dfrac{\omega_n}{\sqrt{\left(1-\frac{c}{m}\omega^{\beta-2}\cos\frac{\beta\pi}{2}\right)}}\sqrt{1-\dfrac{c^2\omega^{2(\beta-1)}\sin^2\frac{\beta\pi}{2}}{4\left(m-c\omega^{\beta-2}\cos\frac{\beta\pi}{2}\right)k}}t\right) \\[4mm] + \dfrac{v_{20} + \dfrac{c\omega^{\beta-1}\sin\frac{\beta\pi}{2}}{2\left(m-c\omega^{\beta-2}\cos\frac{\beta\pi}{2}\right)}x_{20}}{\omega_n\sqrt{1-\dfrac{c^2\omega^{2(\beta-1)}\sin^2\frac{\beta\pi}{2}}{4\left(m-c\omega^{\beta-2}\cos\frac{\beta\pi}{2}\right)k}}} \\[4mm] \times \sin\left(\dfrac{\omega_n}{\sqrt{\left(1-\frac{c}{m}\omega^{\beta-2}\cos\frac{\beta\pi}{2}\right)}}\sqrt{1-\dfrac{c^2\omega^{2(\beta-1)}\sin^2\frac{\beta\pi}{2}}{4\left(m-c\omega^{\beta-2}\cos\frac{\beta\pi}{2}\right)k}}t\right) \end{array} \right]. \tag{7.70}$$

7.3.3.2 Impulse Response of Class II Fractional Vibrators

Denote by $h_2(t)$ the impulse response of a class II fractional vibrator. It is the solution to the fractional differential equation in the form

$$m\dfrac{d^2 h_2(t)}{dt^2} + c\dfrac{d^\beta h_2(t)}{dt^\beta} + kh_2(t) = \delta(t) \tag{7.71}$$

with zero initial conditions. Eq. (7.71) is equivalently expressed by

$$m_{eq2}\dfrac{d^2 h_2(t)}{dt^2} + c_{eq2}\dfrac{dh_2(t)}{dt} + kh_2(t) = \delta(t). \tag{7.72}$$

From (7.72), we have

$$h_2(t) = e^{-\varsigma_{eq2}\omega_{eqn2}t}\frac{1}{m_{eq2}\omega_{eqd2}}\sin\omega_{eqd2}t, \quad t \geq 0. \quad (7.73)$$

Substituting m_{eq2}, ς_{eq2}, ω_{eqn2}, and ω_{eqd2} into (7.73) results in

$$h_2(t) = \frac{e^{-\frac{\varsigma\omega_n\omega^{\beta-1}\sin\frac{\beta\pi}{2}}{1-\frac{c}{m}\omega^{\beta-2}\cos\frac{\beta\pi}{2}}t}\sin\frac{\omega_n\sqrt{1-\frac{\varsigma^2\omega^{2(\beta-1)}\sin^2\frac{\beta\pi}{2}}{1-\frac{c}{m}\omega^{\beta-2}\cos\frac{\beta\pi}{2}}}}{\sqrt{1-\frac{c}{m}\omega^{\beta-2}\cos\frac{\beta\pi}{2}}}t}{\omega_n m\sqrt{1-\frac{c}{m}\omega^{\beta-2}\cos\frac{\beta\pi}{2}}\sqrt{1-\frac{\varsigma^2\omega^{2(\beta-1)}\sin^2\frac{\beta\pi}{2}}{1-\frac{c}{m}\omega^{\beta-2}\cos\frac{\beta\pi}{2}}}}, \quad t \geq 0. \quad (7.74)$$

7.3.3.3 Step Response of Class II Fractional Vibrators

Let $g_2(t)$ be the step response of a class II fractional vibrator. It is the solution to

$$m\frac{d^2 g_2(t)}{dt^2} + c\frac{d^\beta g_2(t)}{dt^\beta} + kg_2(t) = u(t) \quad (7.75)$$

with zero initial conditions. Eq. (7.75) equals to

$$m_{eq2}\frac{d^2 g_2(t)}{dt^2} + c_{eq2}\frac{dg_2(t)}{dt} + kg_2(t) = u(t). \quad (7.76)$$

From (7.76), therefore, $g_2(t)$ is given by

$$g_2(t) = \frac{1}{k}\left[1 - \frac{e^{-\varsigma_{eq2}\omega_{eqn2}t}}{\sqrt{1-\varsigma_{eq2}^2}}\cos\left(\omega_{eqd2}t - \phi_2\right)\right], \quad t \geq 0. \quad (7.77)$$

In (7.77), ϕ_2 is given by

$$\phi_2 = \tan^{-1}\frac{\varsigma_{eq2}}{\sqrt{1-\varsigma_{eq2}^2}}. \quad (7.78)$$

Substituting $\zeta_{eq2}(\omega, \beta)$, $\omega_{eqn2}(\omega, \beta)$, and $\omega_{eqd2}(\omega, \beta)$ into (7.77) and (7.78) yields $g_2(t)$ in the form

$$g_2(t) = \frac{1}{k} \left[1 - \frac{e^{-\frac{\varsigma\omega_n\omega^{\beta-1}\sin\frac{\beta\pi}{2}}{1-\frac{c}{m}\omega^{\beta-2}\cos\frac{\beta\pi}{2}}t}}{\sqrt{1-\frac{\varsigma^2\omega^{2(\beta-1)}\sin^2\frac{\beta\pi}{2}}{1-\frac{c}{m}\omega^{\beta-2}\cos\frac{\beta\pi}{2}}}} \cos\left(\frac{\omega_n}{\sqrt{\left(1-\frac{c}{m}\omega^{\beta-2}\cos\frac{\beta\pi}{2}\right)}} \sqrt{1-\frac{\varsigma^2\omega^{2(\beta-1)}\sin^2\frac{\beta\pi}{2}}{1-\frac{c}{m}\omega^{\beta-2}\cos\frac{\beta\pi}{2}}}\,t - \phi_2 \right) \right],$$ (7.79)

and ϕ_2 given by

$$\phi_2 = \tan^{-1} \frac{\frac{\varsigma\omega^{\beta-1}\sin\frac{\beta\pi}{2}}{\sqrt{1-\frac{c}{m}\omega^{\beta-2}\cos\frac{\beta\pi}{2}}}}{\sqrt{1-\frac{\varsigma^2\omega^{2(\beta-1)}\sin^2\frac{\beta\pi}{2}}{1-\frac{c}{m}\omega^{\beta-2}\cos\frac{\beta\pi}{2}}}}.$$ (7.80)

7.3.4 Frequency Transfer Function of Class II Fractional Vibrators

Denote by $H_2(\omega)$ the frequency transfer function of a class II fractional vibrator. Then,

$$H_2(\omega) = \frac{1/k}{1-\gamma^2\left(1-\frac{c}{m}\omega^{\beta-2}\cos\frac{\beta\pi}{2}\right)+i\dfrac{2\varsigma\omega^\beta\sin\frac{\beta\pi}{2}}{\omega_n}}.$$ (7.81)

Since $\dfrac{c}{m} = 2\varsigma\omega_n$, (7.81) may be rewritten by

$$H_2(\omega) = \frac{1/k}{1-\gamma^2\left(1-2\varsigma\omega_n\omega^{\beta-2}\cos\frac{\beta\pi}{2}\right)+i\dfrac{2\varsigma\omega^\beta\sin\frac{\beta\pi}{2}}{\omega_n}}.$$

Eq. (7.82) is another expression of $H_2(\omega)$. It is in the form

$$H_2(\omega) = \frac{1}{k\left(1 - \gamma_{eq2}^2 + i2\varsigma_{eq2}\gamma_{eq2}\right)}. \tag{7.82}$$

The amplitude-frequency response function is

$$|H_2(\omega)| = \frac{1/k}{\sqrt{\left[1 - \gamma^2\left(1 - \dfrac{c}{m}\omega^{\beta-2}\cos\dfrac{\beta\pi}{2}\right)\right]^2 + \left(\dfrac{2\varsigma\omega^{\beta}}{\omega_n}\sin\dfrac{\beta\pi}{2}\right)^2}}. \tag{7.83}$$

Eq. (7.83) equals to

$$|H_2(\omega)| = \frac{1/k}{\sqrt{\left(1 - \gamma_{eq2}^2\right)^2 + \left(2\varsigma_{eq2}\gamma_{eq2}\right)^2}}. \tag{7.84}$$

Eq. (7.85) is the expression of phase-frequency response function. It is in the form

$$\varphi_2(\omega) = \tan^{-1}\frac{\dfrac{2\varsigma\omega^{\beta}\sin\dfrac{\beta\pi}{2}}{\omega_n}}{1 - \gamma^2\left(1 - \dfrac{c}{m}\omega^{\beta-2}\cos\dfrac{\beta\pi}{2}\right)}. \tag{7.85}$$

For facilitating the computation of $\phi_2(\omega)$, one may use

$$\varphi_2(\omega) = \cos^{-1}\frac{1 - \gamma_{eq2}^2}{\sqrt{\left(1 - \gamma_{eq2}^2\right)^2 + \left(2\varsigma_{eq2}\gamma_{eq2}\right)^2}}. \tag{7.86}$$

7.3.5 Equivalent Logarithmic Decrement and Q Factor of Class II Fractional Vibrators

7.3.5.1 Equivalent Logarithmic Decrement of Class II Fractional Vibrators

Let t_i and t_{i+1} be two successive time points of the free response $x_2(t)$, where $x_2(t)$ reaches its successive peak values of $x_2(t_i)$ and $x_2(t_{i+1})$, respectively. Denote by Δ_{eq2} the equivalent logarithmic decrement of $x_2(t)$. Then,

$$\Delta_{eq2} = \Delta_{eq2}(\omega,\beta) = \frac{2\pi}{\sqrt{1 - \left(\dfrac{\varsigma\omega^{\beta-1}\sin\dfrac{\beta\pi}{2}}{\sqrt{1 - \dfrac{c}{m}\omega^{\beta-2}\cos\dfrac{\beta\pi}{2}}}\right)^2}} \cdot \frac{\varsigma\omega^{\beta-1}\sin\dfrac{\beta\pi}{2}}{\sqrt{1 - \dfrac{c}{m}\omega^{\beta-2}\cos\dfrac{\beta\pi}{2}}}. \tag{7.87}$$

Eq. (7.87) exhibits that Δ_{eq2} is not dimensionless as that of the conventional one. Besides, it is in general not a constant but a function of ω and β. If $\beta = 1$; however, it reduces to the conventional logarithmic decrement. That is,

$$\Delta_{eq2}\big|_{\beta=1} = \frac{2\pi\varsigma}{\sqrt{1-\varsigma^2}}.$$

7.3.5.2 Equivalent Q Factor of Class II Fractional Vibrators

Let Q_{eq2} be the equivalent Q factor of a class II fractional vibrator. Then,

$$Q_{eq2} = Q_{eq2}(\omega,\beta) = \frac{\sqrt{1 - \dfrac{c}{m}\omega^{\beta-2}\cos\dfrac{\beta\pi}{2}}}{2\varsigma\omega^{\beta-1}\sin\dfrac{\beta\pi}{2}}. \tag{7.88}$$

Eq. (7.88) implies that Q_{eq2} is generally not a constant but a function of ω and β. If $\beta = 1$, it reduces to the conventional Q factor $\dfrac{1}{2\varsigma}$. In addition, it is not dimensionless unless $\beta = 1$.

7.4 RESULTS FOR CLASS III FRACTIONAL VIBRATIONS

7.4.1 Equivalent Motion Equation of Class III Fractional Vibrators

Consider the motion equation of a class III fractional vibrator in the form

$$B_3(t) \triangleq m\frac{d^\alpha x_3(t)}{dt^\alpha} + c\frac{d^\beta x_3(t)}{dt^\beta} + kx_3(t) = 0. \tag{7.89}$$

In (7.89), $1 < \alpha < 3$ and $0 < \beta < 2$. It can be equivalently expressed by

$$\begin{aligned} A_3(t) &\triangleq -\left(m\omega^{\alpha-2}\cos\frac{\alpha\pi}{2} + c\omega^{\beta-2}\cos\frac{\beta\pi}{2}\right)\frac{d^2x_3(t)}{dt^2} \\ &+ \left(m\omega^{\alpha-1}\sin\frac{\alpha\pi}{2} + c\omega^{\beta-1}\sin\frac{\beta\pi}{2}\right)\frac{dx_3(t)}{dt} + kx_3(t) = 0. \end{aligned} \tag{7.90}$$

In fact, $F[A_3(t) - B_3(t)] = 0$.

7.4.2 Equivalent Vibration Parameters of Class III Fractional Vibrators

7.4.2.1 Equivalent Mass of Class III Fractional Vibrators

Let m_{eq3} be the equivalent mass for a class III fractional vibrator. Then, from (7.90), we have

$$m_{eq3} = -\left(m\omega^{\alpha-2} \cos\frac{\alpha\pi}{2} + c\omega^{\beta-2} \cos\frac{\beta\pi}{2} \right). \qquad (7.91)$$

Eq. (7.91) implies that the unit of m_{eq3} is not kg unless $\alpha = 2$ and $\beta = 1$. If $\alpha = 2$ and $\beta = 1$, m_{eq3} reduces to the primary mass m. When considering $\frac{c}{m} = 2\varsigma\omega_n$ in (7.91), we have

$$m_{eq3} = m_{eq3}(\omega,\alpha,\beta) = -m\left(\omega^{\alpha-2} \cos\frac{\alpha\pi}{2} + 2\varsigma\omega_n\omega^{\beta-2} \cos\frac{\beta\pi}{2} \right).$$

The asymptotic properties of m_{eq3} are interesting. From (7.91), we have

$$\lim_{\omega\to\infty} m_{eq3}(\omega,\alpha,\beta) = \begin{cases} \infty, 2 < \alpha < 3, \\ 0, 1 < \alpha < 2. \end{cases} \qquad (7.92)$$

In addition, when $1 < \beta < 2$, we have

$$\lim_{\omega\to 0} m_{eq3}(\omega,\alpha,\beta) = \infty. \qquad (7.93)$$

Moreover, if $0 < \beta < 1$, m_{eq3} is negative when $\omega \to 0$.

7.4.2.2 Equivalent Damping of Class III Fractional Vibrators

Let c_{eq3} be the equivalent damping for a class III fractional vibrator. Then, from (7.90), we have

$$c_{eq3} = c_{eq3}(\omega,\alpha,\beta) = m\omega^{\alpha-1} \sin\frac{\alpha\pi}{2} + c\omega^{\beta-1} \sin\frac{\beta\pi}{2}. \qquad (7.94)$$

Eq. (7.94) shows that c_{eq3} is proportional to m and c. From (7.94), we have

$$\lim_{\omega\to\infty} c_{eq3} = \begin{cases} \infty, 1 < \alpha < 2, 0 < \beta < 2, \\ -\infty, 2 < \alpha < 3, 0 < \beta < 1, \end{cases} \qquad (7.95)$$

$$\lim_{\omega \to 0} c_{eq3}(\omega, \alpha, \beta) = 0, \quad 1 < \alpha < 3, 1 < \beta < 2, \tag{7.96}$$

$$\lim_{\omega \to 0} c_{eq3}(\omega, \alpha, \beta) = \infty, \quad 1 < \alpha < 3, 0 < \beta < 1. \tag{7.97}$$

Thus,

$$c_{eq3} \in (-\infty, \infty). \tag{7.98}$$

If $1 < \alpha < 3$ and $0 < \beta < 1$, in the case of large ω, we have

$$c_{eq3} \propto m\omega^{\alpha-1} \sin\frac{\alpha\pi}{2}. \tag{7.99}$$

On the contrary, for $1 < \alpha < 3$ and $0 < \beta < 1$, if ω is small enough,

$$c_{eq3} \propto c\omega^{\beta-1} \sin\frac{\beta\pi}{2}. \tag{7.100}$$

7.4.2.3 Equivalent Damping Ratio of Class III Fractional Vibrators

Let ζ_{eq3} be the equivalent damping ratio for a class III fractional vibrator. Define it by

$$\zeta_{eq3} = \frac{c_{eq3}}{2\sqrt{m_{eq3}k}}. \tag{7.101}$$

Substituting m_{eq3} and c_{eq3} into (7.101) yields

$$\zeta_{eq3} = \zeta_{eq3}(\omega, \alpha, \beta) = \frac{\omega^{\alpha-1}\sin\frac{\alpha\pi}{2} + 2\zeta\omega_n\omega^{\beta-1}\sin\frac{\beta\pi}{2}}{2\omega_n\sqrt{-\left(\omega^{\alpha-2}\cos\frac{\alpha\pi}{2} + 2\zeta\omega_n\omega^{\beta-2}\cos\frac{\beta\pi}{2}\right)}}. \tag{7.102}$$

Eq. (7.102) implies that ζ_{eq3} is dimensionless if and only if $\alpha = 2$ and $\beta = 1$. As a matter of fact, $\zeta_{eq3}(\omega, 2, 1) = \zeta$. If $1 < \alpha < 2$ and $0 < \beta < 2$, we have

$$\zeta_{eq3} \geq 0. \tag{7.103}$$

However, when $2 < \alpha < 3$, it may be negative.

7.4.2.4 Equivalent Damping-Free Natural Frequency of Class III Fractional Vibrators

Let ω_{eqn3} be the equivalent damping-free natural angular frequency for a class III fractional vibrator. Define it by

$$\omega_{eqn3} = \sqrt{\frac{k}{m_{eq3}}}.$$ (7.104)

Substituting m_{eq3} into (7.104) produces

$$\omega_{eqn3} = \frac{\omega_n}{\sqrt{-\left(\omega^{\alpha-2}\cos\frac{\alpha\pi}{2} + \frac{c}{m}\omega^{\beta-2}\cos\frac{\beta\pi}{2}\right)}}.$$ (7.105)

From (7.105), we see that the unit of ω_{eqn3} is generally not rad/s. Its unit reduces to rad/s if $\alpha = 2$ and $\beta = 1$. In fact,

$$\omega_{eqn3}\big|_{\alpha=2,\beta=1} = \omega_n.$$ (7.106)

7.4.2.5 Equivalent Damped Natural Frequency of Class III Fractional Vibrators

Without losing the generality, small damping $|\zeta_{eq3}| \leq 1$ is defaulted in what follows from a view of vibration engineering. Denote the equivalent damped natural angular frequency for a class III fractional vibrator as ω_{eqd3}. Define it by

$$\omega_{eqd3} = \omega_{eqn3}\sqrt{1 - \zeta_{eq3}^2}.$$ (7.107)

Substituting ω_{eqn3} and ζ_{eq3} into (7.107) yields

$$\omega_{eqd3} = \frac{\omega_n\sqrt{1 - \left[\frac{\left(\omega^{\alpha-1}\sin\frac{\alpha\pi}{2} + 2\varsigma\omega_n\omega^{\beta-1}\sin\frac{\beta\pi}{2}\right)^2}{4\omega_n^2\left[-\left(\omega^{\alpha-2}\cos\frac{\alpha\pi}{2} + 2\varsigma\omega_n\omega^{\beta-2}\cos\frac{\beta\pi}{2}\right)\right]}\right]^2}}{\sqrt{-\left(\omega^{\alpha-2}\cos\frac{\alpha\pi}{2} + \frac{c}{m}\omega^{\beta-2}\cos\frac{\beta\pi}{2}\right)}}.$$ (7.108)

From (7.108), we see that ω_{eqd3} degenerates the conventional ω_d with the unit of rad/s when $\alpha = 2$ and $\beta = 1$. In general, its unit is not rad/s.

7.4.2.6 Equivalent Frequency Ratio of Class III Fractional Vibrators

Let γ_{eq3} be the equivalent frequency ratio for a class III fractional vibrator. It is defined by

$$\gamma_{eq3} = \frac{\omega}{\omega_{eqn3}}. \tag{7.109}$$

Substituting ω_{eqn3} into (7.109) produces

$$\gamma_{eq3} = \gamma_{eq3}(\omega,\alpha,\beta) = \gamma \sqrt{-\left(\omega^{\alpha-2}\cos\frac{\alpha\pi}{2} + \frac{c}{m}\omega^{\beta-2}\cos\frac{\beta\pi}{2}\right)}. \tag{7.110}$$

From (7.110), we see that the unit of γ_{eq3} is not dimensionless in general. It reduces to being dimensionless if and only if $\alpha = 2$ and $\beta = 1$. In fact, $\gamma_{eq3}(\omega, 2, 1) = \gamma$.

7.4.3 Responses of Class III Fractional Vibration Systems

7.4.3.1 Free Response of Class III Fractional Vibration Systems

Let $x_3(t)$ be the free response of a class III fractional vibrator. It is the solution to the fractional differential equation given by

$$\left[m\frac{d^\alpha x_3(t)}{dt^\alpha} + c\frac{d^\beta x_3(t)}{dt^\beta} + kx_3(t) = 0, \atop x_3(0) = x_{30}, x_3'(0) = v_{30}. \right. \tag{7.111}$$

The solution to (7.111) is equivalent to the one to the differential equation (7.112) in the form

$$\left[\frac{d^2 x_3(t)}{dt^2} + 2\varsigma_{eq3}\omega_{eqn3}\frac{dx_3(t)}{dt} + \omega_{eqn3}^2 x_3(t) = 0, \atop x_3(0) = x_{30}, x_3'(0) = v_{30}. \right. \tag{7.112}$$

The solution to (7.112) is given by

$$x_3(t) = e^{-\varsigma_{eq3}\omega_{eqn3}t}\left(x_{30}\cos\omega_{eqd3}t + \frac{v_{30} + \varsigma_{eq3}\omega_{eqn3}x_{30}}{\omega_{eqd3}}\sin\omega_{eqd3}t \right), \quad t \geq 0. \tag{7.113}$$

Substituting ζ_{eq3} and ω_{eqn3} into (7.113) yields

$$x_3(t) = e^{-\dfrac{mw^{\alpha-1}\sin\frac{\alpha\pi}{2}+cw^{\beta-1}\sin\frac{\beta\pi}{2}}{2\left(mw^{\alpha-2}\left|\cos\frac{\alpha\pi}{2}\right|-cw^{\beta-2}\cos\frac{\beta\pi}{2}\right)}t}\left(x_{30}\cos\omega_{eqd3}t + A\sin\omega_{eqd3}t\right), \quad t \geq 0. \qquad (7.114)$$

In (7.114), A is given by

$$A = \dfrac{v_{30} + \dfrac{mw^{\alpha-1}\sin\frac{\alpha\pi}{2}+cw^{\beta-1}\sin\frac{\beta\pi}{2}}{2\left(mw^{\alpha-2}\left|\cos\frac{\alpha\pi}{2}\right|-cw^{\beta-2}\cos\frac{\beta\pi}{2}\right)}x_{30}}{\dfrac{\omega_n}{\sqrt{w^{\alpha-2}\left|\cos\frac{\alpha\pi}{2}\right|-\frac{c}{m}w^{\beta-2}\cos\frac{\beta\pi}{2}}}\sqrt{1-\dfrac{\left(mw^{\alpha-1}\sin\frac{\alpha\pi}{2}+cw^{\beta-1}\sin\frac{\beta\pi}{2}\right)^2}{4\left[-\left(mw^{\alpha-2}\cos\frac{\alpha\pi}{2}+cw^{\beta-2}\cos\frac{\beta\pi}{2}\right)k\right]}}}. \qquad (7.115)$$

7.4.3.2 Impulse Response of Class III Fractional Vibrators

Denote by $h_3(t)$ the impulse response of a class III fractional vibrator. It is the solution to the following fractional differential equation

$$m\frac{d^\alpha h_3(t)}{dt^\alpha} + c\frac{d^\beta h_3(t)}{dt^\beta} + kh_3(t) = \delta(t) \qquad (7.116)$$

with zero initial conditions. The equivalent equation of (7.116) is

$$m_{eq3}\frac{d^2 h_3(t)}{dt^2} + c_{eq3}\frac{dh_3(t)}{dt} + kh_3(t) = \delta(t). \qquad (7.117)$$

From (7.117), we have

$$h_3(t) = e^{-\zeta_{eq3}\omega_{eqn3}t}\frac{1}{m_{eq3}\omega_{eqd3}}\sin\omega_{eqd3}t, \quad t \geq 0. \qquad (7.118)$$

Substituting m_{eq3}, ζ_{eq3}, and ω_{eqn3} into (7.118) results in

$$h_3(t) = \dfrac{e^{-\dfrac{mw^{\alpha-1}\sin\frac{\alpha\pi}{2}+cw^{\beta-1}\sin\frac{\beta\pi}{2}}{2\sqrt{-\left(mw^{\alpha-2}\cos\frac{\alpha\pi}{2}+cw^{\beta-2}\cos\frac{\beta\pi}{2}\right)k}}\omega_{eqn3}t}\sin\omega_{eqd3}t}{-\left(mw^{\alpha-2}\cos\frac{\alpha\pi}{2}+cw^{\beta-2}\cos\frac{\beta\pi}{2}\right)\omega_{eqd3}}, \quad t \geq 0. \qquad (7.119)$$

7.4.3.3 Step Response of Class III Fractional Vibrators

Let $g_3(t)$ be the step response of a class III fractional vibrator. It is the solution to

$$m\frac{d^{\alpha}g_3(t)}{dt^{\alpha}}+c\frac{d^{\beta}g_3(t)}{dt^{\beta}}+kg_3(t)=u(t) \tag{7.120}$$

with zero initial conditions. The solution to (7.120) equals to the one to

$$m_{eq3}\frac{d^2 g_3(t)}{dt^2}+c_{eq3}\frac{dg_3(t)}{dt}+kg_3(t)=u(t). \tag{7.121}$$

Since (7.121) is a standard vibration equation in form, we can easily find $g_3(t)$ expressed by

$$g_3(t)=\frac{1}{k}\left[1-\frac{e^{-\varsigma_{eq3}\omega_{eqn3}t}}{\sqrt{1-\varsigma_{eq3}^2}}\cos\left(\omega_{eqd3}t-\phi_3\right)\right], \quad t\geq 0. \tag{7.122}$$

In (7.122),

$$\phi_3=\tan^{-1}\frac{\varsigma_{eq3}}{\sqrt{1-\varsigma_{eq3}^2}}. \tag{7.123}$$

Substitute ς_{eq3} and ω_{eqn3} into (7.122) and (7.123) results in

$$g_3(t)=\frac{1}{k}\left[1-\frac{e^{-\frac{m\omega^{\alpha-1}\sin\frac{\alpha\pi}{2}+c\omega^{\beta-1}\sin\frac{\beta\pi}{2}}{2\sqrt{-\left(m\omega^{\alpha-2}\cos\frac{\alpha\pi}{2}+c\omega^{\beta-2}\cos\frac{\beta\pi}{2}\right)k}}\omega_{eqn3}t}}{\left[1-\frac{m\omega^{\alpha-1}\sin\frac{\alpha\pi}{2}+c\omega^{\beta-1}\sin\frac{\beta\pi}{2}}{2\sqrt{-\left(m\omega^{\alpha-2}\cos\frac{\alpha\pi}{2}+c\omega^{\beta-2}\cos\frac{\beta\pi}{2}\right)k}}\right]^{1/2}}\cos\left(\omega_{eqd3}t-\phi_3\right)\right], \tag{7.124}$$

$$\phi_3 = \tan^{-1} \frac{\dfrac{c\omega^{\beta-1} \sin\dfrac{\beta\pi}{2}}{2\sqrt{\left(m - c\omega^{\beta-2} \cos\dfrac{\beta\pi}{2}\right)k}}}{\sqrt{1 - \dfrac{c^2\omega^{2(\beta-1)} \sin^2\dfrac{\beta\pi}{2}}{4\left(m - c\omega^{\beta-2} \cos\dfrac{\beta\pi}{2}\right)k}}}. \qquad (7.125)$$

7.4.4 Frequency Transfer Function of Class III Fractional Vibration Systems

Denote by $H_3(\omega)$ the frequency transfer function of a class III fractional vibrator. Then,

$$H_3(\omega) = \frac{1/k}{1 - \gamma^2\left(\omega^{\alpha-2}\left|\cos\dfrac{\alpha\pi}{2}\right| - 2\varsigma\omega_n\omega^{\beta-2}\cos\dfrac{\beta\pi}{2}\right) + i\dfrac{\gamma\left(\omega^{\alpha-1}\sin\dfrac{\alpha\pi}{2} + 2\varsigma\omega_n\omega^{\beta-1}\sin\dfrac{\beta\pi}{2}\right)}{\omega_n\left(\omega^{\alpha-2}\left|\cos\dfrac{\alpha\pi}{2}\right| - 2\varsigma\omega_n\omega^{\beta-2}\cos\dfrac{\beta\pi}{2}\right)}}. \qquad (7.126)$$

Since $\dfrac{c}{m} = 2\varsigma\omega_n$, we rewrite (7.126) by

$$H_3(\omega) = \frac{1/k}{1 - \gamma^2\left(\omega^{\alpha-2}\left|\cos\dfrac{\alpha\pi}{2}\right| - \dfrac{c\omega^{\beta-2}\cos\dfrac{\beta\pi}{2}}{m}\right) + i\dfrac{\gamma\left(\omega^{\alpha-1}\sin\dfrac{\alpha\pi}{2} + \dfrac{m}{c}\omega^{\beta-1}\sin\dfrac{\beta\pi}{2}\right)}{\omega_n\left(\omega^{\alpha-2}\left|\cos\dfrac{\alpha\pi}{2}\right| - \dfrac{m}{c}\omega^{\beta-2}\cos\dfrac{\beta\pi}{2}\right)}}. \qquad (7.127)$$

A general form of $H_3(\omega)$ can be written by

$$H_3(\omega) = \frac{1}{k\left(1 - \gamma_{eq3}^2 + i2\varsigma_{eq3}\gamma_{eq3}\right)}. \qquad (7.128)$$

Write $H_3(\omega) = |H_3(\omega)|\exp[-\phi_3(\omega)]$. Then,

$$|H_3(\omega)| = \frac{1/k}{\sqrt{\left(1 - \gamma_{eq3}^2\right)^2 + \left(2\varsigma_{eq3}\gamma_{eq3}\right)^2}}. \qquad (7.129)$$

Expanding (7.129) yields

$$|H_3(\omega)| = \cfrac{1/k}{\sqrt{\left\{ \begin{array}{l} \left[1-\gamma^2\left(\omega^{\alpha-2}\left|\cos\dfrac{\alpha\pi}{2}\right|-2\varsigma\omega_n\omega^{\beta-2}\cos\dfrac{\beta\pi}{2}\right)\right]^2 \\[4mm] +\left[\dfrac{\gamma\left(\omega^{\alpha-1}\sin\dfrac{\alpha\pi}{2}+2\varsigma\omega_n\omega^{\beta-1}\sin\dfrac{\beta\pi}{2}\right)}{\omega_n\left(\omega^{\alpha-2}\left|\cos\dfrac{\alpha\pi}{2}\right|-2\varsigma\omega_n\omega^{\beta-2}\cos\dfrac{\beta\pi}{2}\right)}\right]^2 \end{array} \right\}}}. \qquad (7.130)$$

The phase-frequency response function $\phi_3(\omega)$ is in the form

$$\varphi_3(\omega) = \tan^{-1}\cfrac{\gamma\left(\omega^{\alpha-1}\sin\dfrac{\alpha\pi}{2}+2\varsigma\omega_n\omega^{\beta-1}\sin\dfrac{\beta\pi}{2}\right)}{\cfrac{\omega_n\left(\omega^{\alpha-2}\left|\cos\dfrac{\alpha\pi}{2}\right|-2\varsigma\omega_n\omega^{\beta-2}\cos\dfrac{\beta\pi}{2}\right)}{1-\gamma^2\left(\omega^{\alpha-2}\left|\cos\dfrac{\alpha\pi}{2}\right|-2\varsigma\omega_n\omega^{\beta-2}\cos\dfrac{\beta\pi}{2}\right)}}. \qquad (7.131)$$

When computing $\phi_3(\omega)$ using digital computers, we use

$$\varphi_3(\omega) = \cos^{-1}\cfrac{1-\gamma_{eq3}^2}{\sqrt{\left(1-\gamma_{eq3}^2\right)^2+\left(2\varsigma_{eq3}\gamma_{eq3}\right)^2}}. \qquad (7.132)$$

7.4.5 Equivalent Logarithmic Decrement and Q Factor of Class III Fractional Vibrators

7.4.5.1 Equivalent Logarithmic Decrement of Class III Fractional Vibrators

Suppose that t_i and t_{i+1} are two successive time points of the free response $x_3(t)$, where $x_3(t)$ reaches its successive peak values of $x_3(t_i)$ and $x_3(t_{i+1})$, respectively. Denote by Δ_{eq3} the equivalent logarithmic decrement of $x_3(t)$.

Then,

$$\Delta_{eq3} = \cfrac{\pi\left(\omega^{\alpha-1}\sin\dfrac{\alpha\pi}{2}+2\varsigma w_n\omega^{\beta-1}\sin\dfrac{\beta\pi}{2}\right)}{w_n\sqrt{-\left(\omega^{\alpha-2}\cos\dfrac{\alpha\pi}{2}+2\varsigma w_n\omega^{\beta-2}\cos\dfrac{\beta\pi}{2}\right)}}}{\sqrt{1-\left[\cfrac{\omega^{\alpha-1}\sin\dfrac{\alpha\pi}{2}+2\varsigma w_n\omega^{\beta-1}\sin\dfrac{\beta\pi}{2}}{2w_n\sqrt{-\left(\omega^{\alpha-2}\cos\dfrac{\alpha\pi}{2}+2\varsigma w_n\omega^{\beta-2}\cos\dfrac{\beta\pi}{2}\right)}}\right]^2}}. \qquad (7.133)$$

It can be easily seen from (7.133) that Δ_{eq3}, for $\alpha = 2$ and $\beta = 1$, reduces to the conventional logarithmic decrement $\dfrac{2\pi\varsigma}{\sqrt{1-\varsigma^2}}$.

7.4.5.2 Equivalent Q Factor of Class III Fractional Vibrators

Let Q_{eq3} be the equivalent Q factor of a class III fractional vibrator. Then,

$$Q_{eq3} = Q_{eq3}(\omega,\alpha,\beta) = \cfrac{w_n\sqrt{-\left(\omega^{\alpha-2}\cos\dfrac{\alpha\pi}{2}+2\varsigma w_n\omega^{\beta-2}\cos\dfrac{\beta\pi}{2}\right)}}{\omega^{\alpha-1}\sin\dfrac{\alpha\pi}{2}+2\varsigma w_n\omega^{\beta-1}\sin\dfrac{\beta\pi}{2}}. \qquad (7.134)$$

Eq. (7.134) exhibits that Q_{eq3} is generally not a constant but a function in terms of ω, α, and β. When $\alpha = 2$ and $\beta = 1$, it reduces to the conventional Q factor $\dfrac{1}{2\varsigma}$.

7.5 RESULTS FOR CLASS IV FRACTIONAL VIBRATIONS

7.5.1 Equivalent Motion Equation of Class IV Fractional Vibrators

The motion equation of a class IV fractional vibrator is given by

$$B_4(t) \triangleq m\frac{d^\alpha x_4(t)}{dt^\alpha}+k\frac{d^\lambda x_4(t)}{dt^\lambda}=0, \quad 1<\alpha<3,\ 0\le\lambda<1. \quad (7.135)$$

Eq. (7.136) is the equivalent equation of (7.135). It is expressed by

$$A_4(t) \triangleq -m\omega^{\alpha-2}\cos\frac{\alpha\pi}{2}\frac{d^2 x_4(t)}{dt^2}+\left(m\omega^{\alpha-1}\sin\frac{\alpha\pi}{2}+k\omega^{\lambda-1}\sin\frac{\lambda\pi}{2}\right)\frac{dx_4(t)}{dt}$$

$$+k\omega^\lambda\cos\frac{\lambda\pi}{2}x_4(t)=0. \qquad (7.136)$$

As a matter of fact, $F[A_4(t) - B_4(t)] = 0$.

7.5.2 Equivalent Vibration Parameters of Class IV Fractional Vibration Systems

7.5.2.1 Equivalent Mass of Class IV Fractional Vibrators

Let m_{eq4} be the equivalent mass for a class IV fractional vibrator. From (7.136) and (7.3), we have

$$m_{eq4} = m_{eq1} = -m\omega^{\alpha-2} \cos\frac{\alpha\pi}{2}. \qquad (7.137)$$

7.5.2.2 Equivalent Damping of Class IV Fractional Vibrators

Let c_{eq4} be the equivalent damping for a class IV fractional vibrator. From (7.136), we have

$$c_{eq4} = c_{eq4}(\omega,\alpha,\lambda) = m\omega^{\alpha-1}\sin\frac{\alpha\pi}{2} + k\omega^{\lambda-1}\sin\frac{\lambda\pi}{2}. \qquad (7.138)$$

Eq. (7.138) exhibits that c_{eq4} is proportional to the primary mass m and the primary stiffness k. It is consistent with the function form of the Rayleigh damping. Asymptotically, we have

$$\lim_{\omega\to 0} c_{eq4}(\omega,\alpha,\lambda) = \infty, \quad \lambda \neq 0. \qquad (7.139)$$

Besides,

$$\lim_{\omega\to\infty} c_{eq4}(\omega,\alpha,\lambda) = \begin{cases} \infty, 1 < \alpha < 2, \\ -\infty, 2 < \alpha < 3. \end{cases} \qquad (7.140)$$

Therefore,

$$c_{eq4} \in (-\infty, \infty). \qquad (7.141)$$

Note that $c_{eq4} = 0$ when $\alpha = 2$ and $\lambda = 0$.

7.5.2.3 Equivalent Stiffness of Class IV Fractional Vibrators

Let k_{eq4} be the equivalent stiffness of a class IV fractional vibrator. From (7.136), we see that it is given by

$$k_{eq4} = k_{eq4}(\omega,\lambda) = k\omega^{\lambda}\cos\frac{\lambda\pi}{2}. \qquad (7.142)$$

7.5.2.4 Equivalent Damping Ratio of Class IV Fractional Vibrators

Let ζ_{eq4} be the equivalent damping ratio for a class IV fractional vibrator. Define it by

$$\zeta_{eq4} = \frac{c_{eq4}}{2\sqrt{m_{eq4}k_{eq4}}}. \tag{7.143}$$

Substituting m_{eq4}, c_{eq4}, and k_{eq4} into (7.143) produces

$$\zeta_{eq4} = \zeta_{eq4}(\omega, \alpha, \lambda) = \frac{m\omega^{\alpha-1}\sin\dfrac{\alpha\pi}{2} + k\omega^{\lambda-1}\sin\dfrac{\lambda\pi}{2}}{2\sqrt{mk\omega^{\alpha+\lambda-2}}\left|\cos\dfrac{\alpha\pi}{2}\right|\cos\dfrac{\lambda\pi}{2}}. \tag{7.144}$$

Eq. (7.144) implies that $\zeta_{eq4} = 0$ if $\alpha = 2$ and $\lambda = 0$.

7.5.2.5 Equivalent Damping-Free Natural Frequency of Class IV Fractional Vibrators

Let ω_{eqn4} be the equivalent damping-free natural angular frequency for a class IV fractional vibrator. It is defined by

$$\omega_{eqn4} = \sqrt{\frac{k_{eq4}}{m_{eq4}}}. \tag{7.145}$$

Substituting m_{eq4} and k_{eq4} into (7.145) results in

$$\omega_{eqn4} = \omega_n\sqrt{\frac{\omega^{\lambda}\cos\dfrac{\lambda\pi}{2}}{-\omega^{\alpha-2}\cos\dfrac{\alpha\pi}{2}}}. \tag{7.146}$$

Eq. (7.146) exhibits that the unit of ω_{eqn4} is not rad/s. It deteriorates to rad/s when $\alpha = 2$ and $\lambda = 0$. That is,

$$\omega_{eqn4}\Big|_{\alpha=2,\lambda=0} = \omega_n. \tag{7.147}$$

7.5.2.6 Equivalent Damped Natural Frequency of Class IV Fractional Vibrators

From a view of engineering, small damping $|\varsigma_{eq4}| \leq 1$ is defaulted in what follows. Denote the equivalent damped natural angular frequency for a class IV fractional vibrator as ω_{eqd4}. Define it by

$$\omega_{eqd4} = \omega_{eqn4}\sqrt{1-\varsigma_{eq4}^2}.$$ (7.148)

Substituting ς_{eq4} and ω_{eqn4} into (7.148) produces

$$\omega_{eqd4} = \omega_n \sqrt{\frac{\omega^\lambda \cos\frac{\lambda\pi}{2}}{-\omega^{\alpha-2}\cos\frac{\alpha\pi}{2}}}\sqrt{1 - \left(\frac{m\omega^{\alpha-1}\sin\frac{\alpha\pi}{2} + k\omega^{\lambda-1}\sin\frac{\lambda\pi}{2}}{2\sqrt{mk\omega^{\alpha+\lambda-2}}\left|\cos\frac{\alpha\pi}{2}\right|\cos\frac{\lambda\pi}{2}}\right)^2}.$$ (7.149)

From (7.149), for $\alpha = 2$ and $\lambda = 0$, we have

$$\omega_{eqd4}\big|_{\alpha=2,\lambda=0} = \omega_n.$$ (7.150)

7.5.2.7 Equivalent Frequency Ratio of Class IV Fractional Vibrators

Let γ_{eq4} be the equivalent frequency ratio for a class IV fractional vibrator. It is defined by

$$\gamma_{eq4} = \frac{\omega}{\omega_{eqn4}}.$$ (7.151)

Substituting ω_{eqn4} into (7.151) yields

$$\gamma_{eq4} = \gamma\sqrt{\frac{-\omega^{\alpha-2}\cos\frac{\alpha\pi}{2}}{\omega^\lambda \cos\frac{\lambda\pi}{2}}}.$$ (7.152)

Eq. (7.152) implies that the unit of γ_{eq4} is not dimensionless. It reduces to dimensionless if and only if $\alpha = 2$ and $\lambda = 0$.

7.5.3 Responses of Class IV Fractional Vibration Systems

7.5.3.1 Free Response of Class IV Fractional Vibrators

Let $x_4(t)$ be the free response of a class IV fractional vibrator. It is the solution to the fractional differential equation given by

$$\begin{cases} m\dfrac{d^\alpha x_4(t)}{dt^\alpha} + k\dfrac{d^\lambda x_4(t)}{dt^\lambda} = 0, \\ x_4(0) = x_{40}, x_4'(0) = v_{40}. \end{cases} \tag{7.153}$$

The solution to (7.153) is equivalent to the one to

$$\begin{cases} m_{eq4}\dfrac{d^2 x_4(t)}{dt^2} + c_{eq4}\dfrac{dx_4(t)}{dt} + k_{eq4}x_4(t) = 0, \\ x_4(0) = x_{40}, x_4'(0) = v_{40}. \end{cases} \tag{7.154}$$

The solution to (7.154) is given by

$$x_4(t) = e^{-\varsigma_{eq4}\omega_{eqn4}t}\left(x_{40}\cos\omega_{eqd4}t + \frac{v_{40} + \varsigma_{eq4}\omega_{eqn4}x_{40}}{\omega_{eqd4}}\sin\omega_{eqd4}t\right), \quad t \geq 0. \tag{7.155}$$

Using ς_{eq4} and ω_{eqn4} in the exponential function in (7.155) produces

$$x_4(t) = e^{-\frac{mw^{\alpha-1}\sin\frac{\alpha\pi}{2} + kw^{\lambda-1}\sin\frac{\lambda\pi}{2}}{2\sqrt{mk\omega^{\alpha+\lambda-2}}\left|\cos\frac{\alpha\pi}{2}\right|\cos\frac{\lambda\pi}{2}}\sqrt{\frac{\omega^\lambda\cos\frac{\lambda\pi}{2}}{-\omega^{\alpha-2}\cos\frac{\alpha\pi}{2}}}\omega_n t}$$

$$\left(x_{40}\cos\omega_{eqd4}t + \frac{v_{40} + \varsigma_{eq4}\omega_{eqn4}x_{40}}{\omega_{eqd4}}\sin\omega_{eqd4}t\right), \quad t \geq 0. \tag{7.156}$$

7.5.3.2 Impulse Response of Class IV Fractional Vibration Systems

Denote by $h_4(t)$ the impulse response of a class IV fractional vibrator. It is the solution to the following fractional differential equation

$$m\dfrac{d^\alpha h_4(t)}{dt^\alpha} + k\dfrac{d^\lambda h_4(t)}{dt^\lambda} = \delta(t) \tag{7.157}$$

with zero initial conditions. The equivalent equation of (7.157) is given by

$$\dfrac{d^2 h_4(t)}{dt^2} + 2\varsigma_{eq4}\omega_{eqn4}\dfrac{dh_4(t)}{dt} + \omega_{eqn4}^2 h_4(t) = \dfrac{\delta(t)}{m_{eq4}}. \tag{7.158}$$

The solution to (7.158) is expressed by

$$h_4(t) = e^{-\zeta_{eq4}\omega_{eqn4}t} \frac{1}{m_{eq4}\omega_{eqd4}} \sin \omega_{eqd4}t, \quad t \geq 0. \tag{7.159}$$

Substituting ζ_{eq4} and ω_{eqn4} into the exponential function in (7.159) produces

$$h_4(t) = e^{-\frac{m\omega^{\alpha-1}\sin\frac{\alpha\pi}{2}+k\omega^{\lambda-1}\sin\frac{\lambda\pi}{2}}{2\sqrt{mk\omega^{\alpha+\lambda-2}}\left|\cos\frac{\alpha\pi}{2}\right|\cos\frac{\lambda\pi}{2}}\sqrt{\frac{\omega^{\lambda}\cos\frac{\lambda\pi}{2}}{-\omega^{\alpha-2}\cos\frac{\alpha\pi}{2}}}\omega_n t} \frac{1}{m_{eq4}\omega_{eqd4}} \sin \omega_{eqd4}t, \quad t \geq 0. \tag{7.160}$$

7.5.3.3 Step Response of Class IV Fractional Vibrators

Let $g_4(t)$ be the step response of a class IV fractional vibrator. It is the solution to

$$m_{eq4}\frac{d^\alpha g_4(t)}{dt^\alpha} + k_{eq4}\frac{d^\lambda g_4(t)}{dt^\lambda} = u(t) \tag{7.161}$$

with zero initial conditions. The solution to (7.161) equals to the one to

$$m_{eq4}\frac{d^2 g_4(t)}{dt^2} + c_{eq4}\frac{dg_4(t)}{dt} + kg_4(t) = u(t). \tag{7.162}$$

Since (7.162) is a standard vibration equation in form, we have

$$g_4(t) = \frac{1}{k_{eq4}}\left[1 - \frac{e^{-\zeta_{eq4}\omega_{eqn4}t}}{\sqrt{1-\zeta_{eq4}^2}}\cos\left(\omega_{eqd4}t - \phi_4\right)\right], \quad t \geq 0. \tag{7.163}$$

In (7.163), ϕ_4 is

$$\phi_4 = \tan^{-1}\frac{\zeta_{eq4}}{\sqrt{1-\zeta_{eq4}^2}}. \tag{7.164}$$

The step response can be rewritten by

$$g_4(t) = \frac{1}{k\omega^\lambda \cos\frac{\lambda\pi}{2}}\left[1 - \frac{e^{-\frac{m\omega^{\alpha-1}\sin\frac{\alpha\pi}{2}+k\omega^{\lambda-1}\sin\frac{\lambda\pi}{2}}{2\sqrt{mk\omega^{\alpha+\lambda-2}}\left|\cos\frac{\alpha\pi}{2}\right|\cos\frac{\lambda\pi}{2}}\sqrt{\frac{\omega^{\lambda}\cos\frac{\lambda\pi}{2}}{-\omega^{\alpha-2}\cos\frac{\alpha\pi}{2}}}\omega_n t}}{\sqrt{1-\zeta_{eq4}^2}}\cos\left(\omega_{eqd4}t - \phi_4\right)\right], \quad t \geq 0, \tag{7.165}$$

$$\phi_4(\omega) = \tan^{-1} \frac{\dfrac{m\omega^{\alpha-1}\sin\dfrac{\alpha\pi}{2} + k\omega^{\lambda-1}\sin\dfrac{\lambda\pi}{2}}{2\sqrt{mk\omega^{\alpha+\lambda-2}}\left|\cos\dfrac{\alpha\pi}{2}\right|\cos\dfrac{\lambda\pi}{2}}}{\sqrt{1 - \left(\dfrac{m\omega^{\alpha-1}\sin\dfrac{\alpha\pi}{2} + k\omega^{\lambda-1}\sin\dfrac{\lambda\pi}{2}}{2\sqrt{mk\omega^{\alpha+\lambda-2}}\left|\cos\dfrac{\alpha\pi}{2}\right|\cos\dfrac{\lambda\pi}{2}}\right)^2}}. \tag{7.166}$$

7.5.4 Frequency Transfer Function of Class IV Fractional Vibrators

Denote by $H_4(\omega)$ the frequency transfer function of a class IV fractional vibrator. It is expressed by

$$H_4(\omega) = \frac{1}{k_{eq4}\left(1 - \gamma_{eq4}^2 + i2\varsigma_{eq4}\gamma_{eq4}\right)}. \tag{7.167}$$

Substituting ζ_{eq4}, k_{eq4}, and γ_{eq4} into (7.167) produces

$$H_4(\omega) = \frac{1}{k\omega^\lambda\cos\dfrac{\lambda\pi}{2}\left[1 - \gamma^2\dfrac{-\omega^{\alpha-2}\cos\dfrac{\alpha\pi}{2}}{\omega^\lambda\cos\dfrac{\lambda\pi}{2}} + i2\gamma\dfrac{m\omega^{\alpha-1}\sin\dfrac{\alpha\pi}{2} + k\omega^{\lambda-1}\sin\dfrac{\lambda\pi}{2}}{2\sqrt{mk\omega^{\alpha+\lambda-2}}\left|\cos\dfrac{\alpha\pi}{2}\right|\cos\dfrac{\lambda\pi}{2}}\sqrt{\dfrac{-\omega^{\alpha-2}\cos\dfrac{\alpha\pi}{2}}{\omega^\lambda\cos\dfrac{\lambda\pi}{2}}}\right]}. \tag{7.168}$$

When writing $H_4(\omega)$ by $H_4(\omega) = |H_4(\omega)|\exp[-\phi_4(\omega)]$, we have

$$\left|H_4(\omega)\right| = \frac{1}{k_{eq4}}\frac{1}{\sqrt{\left(1 - \gamma_{eq4}^2\right)^2 + \left(2\varsigma_{eq4}\gamma_{eq4}\right)^2}}, \tag{7.169}$$

$$\varphi_4(\omega) = \cos^{-1}\frac{1 - \gamma_{eq4}^2}{\sqrt{\left(1 - \gamma_{eq4}^2\right)^2 + \left(2\varsigma_{eq4}\gamma_{eq4}\right)^2}}. \tag{7.170}$$

7.5.5 Equivalent Logarithmic Decrement and Q Factor of Class IV Fractional Vibrators

7.5.5.1 Equivalent Logarithmic Decrement of Class IV Fractional Vibrators

Denote by t_i and t_{i+1} two time points of the free response $x_4(t)$, where $x_4(t_i)$ and $x_4(t_{i+1})$ are its successive peak values at t_i and t_{i+1}. Designate Δ_{eq4} as the equivalent logarithmic decrement of $x_4(t)$. Then,

$$\Delta_{eq4} = \cfrac{2\pi \cfrac{m\omega^{\alpha-1} \sin\dfrac{\alpha\pi}{2} + k\omega^{\lambda-1} \sin\dfrac{\lambda\pi}{2}}{2\sqrt{mk\omega^{\alpha+\lambda-2}} \left|\cos\dfrac{\alpha\pi}{2}\right| \cos\dfrac{\lambda\pi}{2}}}{\sqrt{1 - \left(\cfrac{m\omega^{\alpha-1} \sin\dfrac{\alpha\pi}{2} + k\omega^{\lambda-1} \sin\dfrac{\lambda\pi}{2}}{2\sqrt{mk\omega^{\alpha+\lambda-2}} \left|\cos\dfrac{\alpha\pi}{2}\right| \cos\dfrac{\lambda\pi}{2}}\right)^2}}. \tag{7.171}$$

Eq. (7.171) exhibits that Δ_{eq4} is not dimensionless. If $\alpha = 2$ and $\lambda = 0$, it reduces to zero because the vibration system reduces to a conventional damping-free one. Hence,

$$\Delta_{eq4}\big|_{\alpha=2,\lambda=0} = 0. \tag{7.172}$$

7.5.5.2 Equivalent Q Factor of Class IV Fractional Vibrators

Denote by Q_{eq4} the equivalent Q factor of a class IV fractional vibrator. It is represented by

$$Q_{eq4} = \cfrac{\sqrt{mk\omega^{\alpha+\lambda-2}} \left|\cos\dfrac{\alpha\pi}{2}\right| \cos\dfrac{\lambda\pi}{2}}{m\omega^{\alpha-1} \sin\dfrac{\alpha\pi}{2} + k\omega^{\lambda-1} \sin\dfrac{\lambda\pi}{2}}. \tag{7.173}$$

Eq. (7.173) implies that Q_{eq4} is not a constant but a function in terms of ω, α, and λ. In addition, it is not dimensionless. However, $Q_{eq4} = \infty$ if $\alpha = 2$ and $\lambda = 0$.

7.6 RESULTS FOR CLASS V FRACTIONAL VIBRATORS

7.6.1 Equivalent Motion Equation of Class V Fractional Vibrators

Eq. (7.174) is the motion equation of a class V fractional vibrator. It is in the form

$$B_5(t) \triangleq m\frac{d^2 x_5(t)}{dt^2} + k\frac{d^\lambda x_5(t)}{dt^\lambda} = 0. \tag{7.174}$$

Eq. (7.174) can be equivalently expressed by

$$A_5(t) \triangleq m\frac{d^2 x_5(t)}{dt^2} + k\omega^{\lambda-1}\sin\frac{\lambda\pi}{2}\frac{dx_5(t)}{dt} + k\omega^\lambda \cos\frac{\lambda\pi}{2}x_5(t) = 0. \tag{7.175}$$

In fact, $F[A_5(t) - B_5(t)] = 0$.

7.6.2 Equivalent Vibration Parameters of Class V Fractional Vibrators

7.6.2.1 Equivalent Mass of Class V Fractional Vibrators

Let m_{eq5} be the equivalent mass for a class V fractional vibrator. From (7.175), it is given by

$$m_{eq5} = m. \tag{7.176}$$

7.6.2.2 Equivalent Damping of Class V Fractional Vibrators

Denote by c_{eq5} the equivalent damping of a class V fractional vibrator. From (7.175), it is expressed by

$$c_{eq5} = c_{eq5}(\omega,\lambda) = k\omega^{\lambda-1}\sin\frac{\lambda\pi}{2}. \tag{7.177}$$

Eq. (7.177) exhibits that c_{eq5} is proportional to k with the coefficient $\omega^{\lambda-1}\sin\frac{\lambda\pi}{2}$. It has the asymptotic properties described by

$$\lim_{\omega\to 0} c_{eq5}(\omega,\lambda) = \infty, \tag{7.178}$$

$$\lim_{\omega\to\infty} c_{eq5}(\omega,\lambda) = 0. \tag{7.179}$$

If $\lambda = 0$, $c_{eq5} = 0$. In fact, c_{eq5} is simply produced by fractional displacement.

7.6.2.3 Equivalent Stiffness of Class V Fractional Vibrators

Let k_{eq5} be the equivalent stiffness of a class V fractional vibrator. Then,

$$k_{eq5} = k_{eq4}. \tag{7.180}$$

Eq. (7.180) is true as can be seen from (7.175) and (7.142).

7.6.2.4 Equivalent Damping Ratio of Class V Fractional Vibrators

Let ζ_{eq5} be the equivalent damping ratio for a class V fractional vibrator. Define it by

$$\zeta_{eq5} = \frac{c_{eq5}}{2\sqrt{mk_{eq5}}}. \tag{7.181}$$

Substituting c_{eq5} and k_{eq5} into (7.181) results in

$$\zeta_{eq5} = \zeta_{eq5}(\omega, \lambda) = \frac{k\omega^{\lambda-1} \sin\frac{\lambda\pi}{2}}{2\sqrt{mk\omega^{\lambda} \cos\frac{\lambda\pi}{2}}} = \frac{\omega_n \omega^{\frac{\lambda}{2}-1} \sin\frac{\lambda\pi}{2}}{2\sqrt{\cos\frac{\lambda\pi}{2}}}. \tag{7.182}$$

From (7.182), we see that the quantity ζ_{eq5} is not dimensionless in general. It is a positive quantity. If $\lambda = 0$, $\zeta_{eq5} = 0$.

7.6.2.5 Equivalent Damping-Free Natural Frequency of Class V Fractional Vibrators

For a class V fractional vibrator, denote its equivalent damping-free natural angular frequency as ω_{eqn5}. Define it by

$$\omega_{eqn5} = \sqrt{\frac{k_{eq5}}{m}}. \tag{7.183}$$

Substituting k_{eq5} into (7.183) yields

$$\omega_{eqn5} = \omega_n \sqrt{\omega^{\lambda} \cos\frac{\lambda\pi}{2}}. \tag{7.184}$$

Eq. (7.184) exhibits that the unit of ω_{eqn5} is not rad/s. However, it deteriorates to rad/s if $\lambda = 0$. In fact,

$$\omega_{eqn5}\big|_{\lambda=0} = \omega_n. \tag{7.185}$$

7.6.2.6 Equivalent Damped Natural Frequency of Class V Fractional Vibrators

We default small damping $|\zeta_{eq5}| \leq 1$ in what follows. Let ω_{eqd5} be the equivalent damped natural angular frequency for a class V fractional vibrator. Define it by

$$\omega_{eqd5} = \omega_{eqn5} \sqrt{1 - \zeta_{eq5}^2}. \tag{7.186}$$

Substituting ζ_{eq5} into (7.186) produces

$$\omega_{eqd5} = \omega_n \sqrt{\omega^\lambda \cos \frac{\lambda\pi}{2}} \sqrt{1 - \left(\frac{k\omega^{\lambda-1} \sin \frac{\lambda\pi}{2}}{2\sqrt{mk\omega^\lambda \cos \frac{\lambda\pi}{2}}} \right)^2}. \tag{7.187}$$

Eq. (7.187) implies that the unit of ω_{eqd5} is not rad/s. When $\lambda = 0$, its unit is rad/s. As a matter of fact,

$$\omega_{eqd5}\big|_{\lambda=0} = \omega_n. \tag{7.188}$$

7.6.2.7 Equivalent Frequency Ratio of Class V Fractional Vibrators

Let γ_{eq5} be the equivalent frequency ratio for a class V fractional vibrator. Define it by

$$\gamma_{eq5} = \frac{\omega}{\omega_{eqn5}}. \tag{7.189}$$

Substituting ω_{eqn5} into (7.189) yields

$$\gamma_{eq5} = \gamma \sqrt{\frac{1}{\omega^\lambda \cos \frac{\lambda\pi}{2}}}. \tag{7.190}$$

From (7.190), we see that the unit of γ_{eq5} is not dimensionless. It reduces to being dimensionless if $\lambda = 0$.

7.6.3 Responses of Class V Fractional Vibrators
7.6.3.1 Free Response of Class V Fractional Vibrators
Let $x_5(t)$ be the free response of a class V fractional vibrator. It is the solution to the fractional differential equation in the form

$$\left| m\frac{d^2x_5(t)}{dt^2}+k\frac{d^\lambda x_5(t)}{dt^\lambda}=0, \right.$$
$$x_5(0)=x_{50},x_5'(0)=v_{50}.$$
(7.191)

The equivalent equation of (7.191) is given by

$$\left| m\frac{d^2x_5(t)}{dt^2}+c_{eq5}\frac{dx_5(t)}{dt}+k_{eq5}x_5(t)=0, \right.$$
$$x_5(0)=x_{50},x_5'(0)=v_{50}.$$
(7.192)

From (7.192), we have

$$x_5(t)=e^{-\varsigma_{eq5}\omega_{eqn5}t}\left(x_{50}\cos\omega_{eqd5}t+\frac{v_{50}+\varsigma_{eq5}\omega_{eqn5}x_{50}}{\omega_{eqd5}}\sin\omega_{eqd5}t\right),\quad t\geq 0.$$
(7.193)

7.6.3.2 Impulse Response of Class V Fractional Vibrators
Denote by $h_5(t)$ the impulse response of a class V fractional vibrator. It is the solution to

$$m\frac{d^2h_5(t)}{dt^2}+k\frac{d^\lambda h_5(t)}{dt^\lambda}=\delta(t)$$
(7.194)

with zero initial conditions. Since (7.195) is the equivalent equation of (7.194),

$$m\frac{d^2h_5(t)}{dt^2}+c_{eq5}\frac{dh_5(t)}{dt}+k_{eq5}h_5(t)=\delta(t),$$
(7.195)

we have

$$h_5(t)=e^{-\varsigma_{eq5}\omega_{eqn5}t}\frac{1}{m\omega_{eqd5}}\sin\omega_{eqd5}t,\quad t\geq 0.$$
(7.196)

7.6.3.3 Step Response of Class V Fractional Vibrators

Let $g_5(t)$ be the step response of a class V fractional vibrator. It is the solution to

$$m\frac{d^2 g_5(t)}{dt^2} + k\frac{d^\lambda g_5(t)}{dt^\lambda} = u(t) \qquad (7.197)$$

with zero initial conditions. Because (7.197) equals to

$$m\frac{d^2 g_5(t)}{dt^2} + c_{eq5}\frac{dg_5(t)}{dt} + k_{eq5}g_5(t) = u(t), \qquad (7.198)$$

we have $g_5(t)$ in the form

$$g_5(t) = \frac{1}{k_{eq5}}\left[1 - \frac{e^{-\zeta_{eq5}\omega_{eqn5}t}}{\sqrt{1-\zeta_{eq5}^2}}\cos\left(\omega_{eqd5}t - \phi_5\right)\right], \quad t \geq 0. \qquad (7.199)$$

The phase ϕ_5 is given by

$$\phi_5 = \tan^{-1}\frac{\zeta_{eq5}}{\sqrt{1-\zeta_{eq5}^2}}. \qquad (7.200)$$

7.6.4 Frequency Transfer Function of Class V Fractional Vibrators

Let $H_5(\omega)$ be the frequency transfer function of a class V fractional vibrator. Then,

$$H_5(\omega) = \frac{1}{k_{eq5}\left(1 - \gamma_{eq5}^2 + i2\zeta_{eq5}\gamma_{eq5}\right)}. \qquad (7.201)$$

Substituting ζ_{eq5}, k_{eq5}, and γ_{eq5} into (7.201) produces

$$H_5(\omega) = \frac{1}{k\omega^\lambda \cos\frac{\lambda\pi}{2}\left[1 - \frac{\gamma^2}{\omega^\lambda\cos\frac{\lambda\pi}{2}} + i2\gamma\frac{k\omega^{\lambda-1}\sin\frac{\lambda\pi}{2}}{2\sqrt{mk\omega^\lambda\cos\frac{\lambda\pi}{2}}}\sqrt{\frac{1}{\omega^\lambda\cos\frac{\lambda\pi}{2}}}\right]}. \qquad (7.202)$$

When using $H_5(\omega) = |H_5(\omega)|\exp(-\phi_5(\omega))$, we have

$$|H_5(\omega)| = \frac{1}{k_{eq5}} \frac{1}{\sqrt{\left(1-\gamma_{eq5}^2\right)^2 + \left(2\varsigma_{eq5}\gamma_{eq5}\right)^2}}, \qquad (7.203)$$

$$\varphi_5(\omega) = \cos^{-1} \frac{1-\gamma_{eq5}^2}{\sqrt{\left(1-\gamma_{eq5}^2\right)^2 + \left(2\varsigma_{eq5}\gamma_{eq5}\right)^2}}. \qquad (7.204)$$

7.6.5 Equivalent Logarithmic Decrement and Q Factor of Class V Fractional Vibrators

7.6.5.1 Equivalent Logarithmic Decrement of Class V Fractional Vibrators

Let t_i and t_{i+1} be two time points of the free response $x_5(t)$ so that $x_5(t_i)$ and $x_5(t_{i+1})$ are its successive peak values at t_i and t_{i+1}. Denote by Δ_{eq5} the equivalent logarithmic decrement of $x_5(t)$. Then,

$$\Delta_{eq5} = \frac{2\pi \dfrac{k\omega^{\lambda-1}\sin\dfrac{\lambda\pi}{2}}{2\sqrt{mk\omega^\lambda}\cos\dfrac{\lambda\pi}{2}}}{\sqrt{1 - \left(\dfrac{k\omega^{\lambda-1}\sin\dfrac{\lambda\pi}{2}}{2\sqrt{mk\omega^\lambda}\cos\dfrac{\lambda\pi}{2}}\right)^2}}. \qquad (7.205)$$

7.6.5.2 Equivalent Q Factor of Class V Fractional Vibrators

Let Q_{eq5} be the equivalent Q factor of a class V fractional vibrator. Then,

$$Q_{eq5} = \frac{\sqrt{mk\omega^\lambda}\cos\dfrac{\lambda\pi}{2}}{k\omega^{\lambda-1}\sin\dfrac{\lambda\pi}{2}}. \qquad (7.206)$$

7.7 RESULTS FOR CLASS VI FRACTIONAL VIBRATORS

7.7.1 Equivalent Motion Equation of Class VI Fractional Vibrators

The motion equation of a class VI fractional vibrator is given by

$$B_6(t) \triangleq m\frac{d^\alpha x_6(t)}{dt^\alpha} + c\frac{d^\beta x_6(t)}{dt^\beta} + k\frac{d^\lambda x_6(t)}{dt^\lambda} = 0. \qquad (7.207)$$

Eq. (7.207) is equivalently given by

$$A_6(t) \triangleq -\left(m\omega^{\alpha-2}\cos\frac{\alpha\pi}{2}+c\omega^{\beta-2}\cos\frac{\beta\pi}{2}\right)\frac{d^2x_6(t)}{dt^2}$$

$$+\left(m\omega^{\alpha-1}\sin\frac{\alpha\pi}{2}+c\omega^{\beta-1}\sin\frac{\beta\pi}{2}+k\omega^{\lambda-1}\sin\frac{\lambda\pi}{2}\right)\frac{dx_6(t)}{dt} \quad (7.208)$$

$$+k\omega^\lambda\cos\frac{\lambda\pi}{2}x_6(t)=0.$$

In fact, $F[A_6(t) - B_6(t)] = 0$.

7.7.2 Equivalent Vibration Parameters of Class VI Fractional Vibrators

7.7.2.1 Equivalent Mass of Class VI Fractional Vibrator

Let m_{eq6} be the equivalent mass for a class VI fractional vibrator. From (7.208) and (7.91), we have

$$m_{eq6} = m_{eq3} = -\left(m\omega^{\alpha-2}\cos\frac{\alpha\pi}{2}+c\omega^{\beta-2}\cos\frac{\beta\pi}{2}\right).$$

Note that $m_{eq6} = m_{eq3}$.

7.7.2.2 Equivalent Damping of Class VI Fractional Vibrators

Denote by c_{eq6} the equivalent damping for a class VI fractional vibrator. From (7.208), we have

$$c_{eq6} = c_{eq6}(\omega,\alpha,\beta,\lambda) = m\omega^{\alpha-1}\sin\frac{\alpha\pi}{2}+c\omega^{\beta-1}\sin\frac{\beta\pi}{2}+k\omega^{\lambda-1}\sin\frac{\lambda\pi}{2}. \quad (7.209)$$

If $\omega \to \infty$, for $0 \le \lambda < 1$,

$$c_{eq6} = \begin{cases} \infty, 1<\alpha<2, 0<\beta<2, \\ -\infty, 2<\alpha<3, 0<\beta<1. \end{cases} \quad (7.210)$$

From (7.209), we see that c_{eq6} is proportional to m, c, and k. Asymptotically, c_{eq6} for $\omega \to 0$ is given by

$$\lim_{\omega\to 0} c_{eq6} = \infty. \quad (7.211)$$

For $1 < \alpha < 3$ and $0 < \beta < 1$, when ω is large, we have

$$c_{eq6} \propto m\omega^{\alpha-1}\sin\frac{\alpha\pi}{2}. \quad (7.212)$$

When $1 < \alpha < 3$ and $0 < \beta < 1$, if ω is small enough,

$$c_{eq6} \propto c\omega^{\beta-1} \sin\frac{\beta\pi}{2} + k\omega^{\lambda-1} \sin\frac{\lambda\pi}{2}. \tag{7.213}$$

From this, we infer that $-\infty < c_{eq6} < \infty$.

7.7.2.3 Equivalent Stiffness of Class VI Fractional Vibrators

Let k_{eq6} be the equivalent stiffness of a class VI fractional vibrator. From (7.208) and (7.142), we have

$$k_{eq6} = k_{eq4} = k\omega^{\lambda} \cos\frac{\lambda\pi}{2}.$$

7.7.2.4 Equivalent Damping Ratio of Class VI Fractional Vibrators

Let ζ_{eq6} be the equivalent damping ratio for a class VI fractional vibrator. Define it by

$$\zeta_{eq6} = \frac{c_{eq6}}{2\sqrt{m_{eq6}k_{eq6}}}. \tag{7.214}$$

Substituting m_{eq6}, c_{eq6}, and k_{eq6} into (7.214) produces

$$\zeta_{eq6} = \zeta_{eq6}(\omega, \alpha, \beta, \lambda) = \frac{m\omega^{\alpha-1} \sin\frac{\alpha\pi}{2} + c\omega^{\beta-1} \sin\frac{\beta\pi}{2} + k\omega^{\lambda-1} \sin\frac{\lambda\pi}{2}}{2\sqrt{-\left(m\omega^{\alpha-2} \cos\frac{\alpha\pi}{2} + c\omega^{\beta-2} \cos\frac{\beta\pi}{2}\right)k\omega^{\lambda} \cos\frac{\lambda\pi}{2}}}. \tag{7.215}$$

Eq. (7.215) exhibits that ζ_{eq6} is generally not dimensionless. It is dimensionless if $\alpha = 2$, $\beta = 1$, and $\lambda = 0$. In fact, $\zeta_{eq6}(\omega, 2, 1, 0) = \zeta$. It may be negative relying on values of α, β, and λ.

7.7.2.5 Equivalent Damping-Free Natural Frequency of Class VI Fractional Vibrators

Let ω_{eqn6} be the equivalent damping-free natural angular frequency of a class VI fractional vibrator. It is defined by

$$\omega_{eqn6} = \sqrt{\frac{k_{eq6}}{m_{eq6}}}. \tag{7.216}$$

Substituting m_{eq6} and k_{eq6} into (7.216) yields

$$\omega_{eqn6} = \sqrt{\frac{kw^{\lambda}\cos\dfrac{\lambda\pi}{2}}{-\left(mw^{\alpha-2}\cos\dfrac{\alpha\pi}{2}+cw^{\beta-2}\cos\dfrac{\beta\pi}{2}\right)}}. \qquad (7.217)$$

From (7.217), we see that the unit of ω_{eqn6} is not rad/s in general. It reduces to rad/s if $\alpha = 2$, $\beta = 1$, and $\lambda = 0$. To be precise,

$$\left.\omega_{eqn6}\right|_{\alpha=2,\beta=1,\lambda=0} = \omega_n. \qquad (7.218)$$

7.7.2.6 Equivalent Damped Natural Frequency of Class VI Fractional Vibrators

Denote by ω_{eqd6} the equivalent damped natural angular frequency for a class VI fractional vibrator. Suppose small damping $|\zeta_{eq6}| \le 1$ in what follows. Define ω_{eqd6} by

$$\omega_{eqd6} = \omega_{eqn6}\sqrt{1-\zeta_{eq6}^2}. \qquad (7.219)$$

Substituting ω_{eqn6} and ζ_{eq6} into (7.219) results in

$$\omega_{eqd6} = \sqrt{\frac{kw^{\lambda}\cos\dfrac{\lambda\pi}{2}}{-\left(mw^{\alpha-2}\cos\dfrac{\alpha\pi}{2}+cw^{\beta-2}\cos\dfrac{\beta\pi}{2}\right)}}$$
$$\sqrt{1-\left(\frac{mw^{\alpha-1}\sin\dfrac{\alpha\pi}{2}+cw^{\beta-1}\sin\dfrac{\beta\pi}{2}+kw^{\lambda-1}\sin\dfrac{\lambda\pi}{2}}{2\sqrt{-\left(mw^{\alpha-2}\cos\dfrac{\alpha\pi}{2}+cw^{\beta-2}\cos\dfrac{\beta\pi}{2}\right)kw^{\lambda}\cos\dfrac{\lambda\pi}{2}}}\right)^2}. \qquad (7.220)$$

Eq. (7.220) exhibits that, if $\alpha = 2$, $\beta = 1$, and $\lambda = 0$, ω_{eqd6} degenerates the conventional ω_d with the unit of rad/s. Generally, its unit is not rad/s.

7.7.2.7 Equivalent Frequency Ratio of Class VI Fractional Vibrators
Denote the equivalent frequency ratio for a class VI fractional vibrator by γ_{eq6} and define it by

$$\gamma_{eq6} = \frac{\omega}{\omega_{eqn6}}. \tag{7.221}$$

Substituting ω_{eqn6} into (7.221) produces

$$\gamma_{eq6} = \gamma \sqrt{\frac{-\left(\omega^{\alpha-2}\cos\frac{\alpha\pi}{2} + 2\varsigma\omega_n\omega^{\beta-2}\cos\frac{\beta\pi}{2}\right)}{\omega^{\lambda}\cos\frac{\lambda\pi}{2}}}. \tag{7.222}$$

We see that, from (7.222), γ_{eq6} reduces to γ when $\alpha = 2$, $\beta = 1$, $\lambda = 0$.

7.7.3 Responses of Class VI Vibrators

7.7.3.1 Free Response of Class VI Fractional Vibration Systems
Let $x_6(t)$ be the free response of a class VI fractional vibrator. It is the solution to the fractional differential equation in the form

$$\begin{cases} m\dfrac{d^{\alpha}x_6(t)}{dt^{\alpha}} + c\dfrac{d^{\beta}x_6(t)}{dt^{\beta}} + k\dfrac{d^{\lambda}x_6(t)}{dt^{\lambda}} = 0, \\ x_6(0) = x_{60}, x_6'(0) = v_{60}. \end{cases} \tag{7.223}$$

Eq. (7.224) is the equivalence of (7.223). It is expressed by

$$\begin{cases} m_{eq6}\dfrac{d^2 x_6(t)}{dt^2} + c_{eq6}\dfrac{dx_6(t)}{dt} + k_{eq6}x_6(t) = 0, \\ x_6(0) = x_{60}, x_6'(0) = v_{60}. \end{cases} \tag{7.224}$$

From (7.224), we have

$$x_6(t) = e^{-\varsigma_{eq6}\omega_{eqn6}t}\left(x_{60}\cos\omega_{eqd6}t + \frac{v_{60} + \varsigma_{eq6}\omega_{eqn6}x_{60}}{\omega_{eqd6}}\sin\omega_{eqd6}t\right), \quad t \geq 0. \tag{7.225}$$

7.7.3.2 Impulse Response of Class VI Fractional Vibration Systems

Let $h_6(t)$ be the impulse response of a class VI fractional vibrator. It is the solution to

$$m\frac{d^\alpha h_6(t)}{dt^\alpha} + c\frac{d^\beta h_6(t)}{dt^\beta} + k\frac{d^\lambda h_6(t)}{dt^\lambda} = \delta(t) \qquad (7.226)$$

with zero initial conditions. Because (7.226) can equivalently be expressed by

$$\frac{d^2 h_6(t)}{dt^2} + 2\varsigma_{eq6}\omega_{eqn6}\frac{dh_6(t)}{dt} + \omega_{eqn6}^2 h_6(t) = \frac{\delta(t)}{m_{eq6}}, \qquad (7.227)$$

we have

$$h_6(t) = e^{-\varsigma_{eq6}\omega_{eqn6}t}\frac{1}{m_{eq6}\omega_{eqd6}}\sin\omega_{eqd6}t, \quad t \geq 0. \qquad (7.228)$$

7.7.3.3 Step Response of Class VI Fractional Vibration Systems

Denote by $g_6(t)$ the step response of a class VI fractional vibrator. It is the solution to

$$m\frac{d^\alpha g_6(t)}{dt^\alpha} + c\frac{d^\beta g_6(t)}{dt^\beta} + k\frac{d^\lambda g_6(t)}{dt^\lambda} = u(t) \qquad (7.229)$$

with zero initial conditions. Eq. (7.230) is the equivalence of (7.229). It is in the form

$$m_{eq6}\frac{d^2 g_6(t)}{dt^2} + c_{eq6}\frac{dg_6(t)}{dt} + k_{eq6}g_6(t) = u(t). \qquad (7.230)$$

Thus, from (7.230), we write $g_6(t)$ by

$$g_6(t) = \frac{1}{k_{eq6}}\left[1 - \frac{e^{-\varsigma_{eq6}\omega_{eqn6}t}}{\sqrt{1-\varsigma_{eq6}^2}}\cos\left(\omega_{eqd6}t - \phi_6\right)\right], \quad t \geq 0. \qquad (7.231)$$

In (7.231),

$$\phi_6 = \tan^{-1} \frac{\varsigma_{eq6}}{\sqrt{1-\zeta_{eq6}^2}}.$$ (7.232)

7.7.4 Frequency Transfer Function of Class VI Fractional Vibrators

Let $H_6(\omega)$ be the frequency transfer function of a class VI fractional vibrator. It is expressed by

$$H_6(\omega) = \frac{1}{k\left[\begin{array}{l}\left[\omega^\lambda \cos\frac{\lambda\pi}{2} + \gamma^2\left(\omega^{\alpha-2}\cos\frac{\alpha\pi}{2}+2\varsigma\omega_n\omega^{\beta-2}\cos\frac{\beta\pi}{2}\right)\right] \\ +i\gamma\left(\omega^{\alpha-1}\sin\frac{\alpha\pi}{2}+2\varsigma\omega_n\omega^{\beta-1}\sin\frac{\beta\pi}{2}+\omega_n^2\omega^{\lambda-1}\sin\frac{\lambda\pi}{2}\right)\end{array}\right]}.$$ (7.233)

If writing $H_6(\omega) = |H_6(\omega)|\exp(-\phi_6(\omega))$, we have

$$|H_6(\omega)| = \frac{1}{k_{eq6}} \frac{1}{\sqrt{\left(1-\gamma_{eq6}^2\right)^2 + \left(2\varsigma_{eq6}\gamma_{eq6}\right)^2}}$$
$$= \frac{1}{k} \frac{1}{\sqrt{\left[\omega^\lambda\cos\frac{\lambda\pi}{2}+\gamma^2\left(\omega^{\alpha-2}\cos\frac{\alpha\pi}{2}+2\varsigma\omega_n\omega^{\beta-2}\cos\frac{\beta\pi}{2}\right)\right]^2 + \gamma^2\left(\omega^{\alpha-1}\sin\frac{\alpha\pi}{2}+2\varsigma\omega_n\omega^{\beta-1}\sin\frac{\beta\pi}{2}+\omega_n^2\omega^{\lambda-1}\sin\frac{\lambda\pi}{2}\right)^2}},$$ (7.234)

and

$$\varphi_6(\omega) = \cos^{-1}\frac{1-\gamma_{eq6}^2}{\sqrt{\left(1-\gamma_{eq6}^2\right)^2+\left(2\varsigma_{eq6}\gamma_{eq6}\right)^2}}.$$ (7.235)

7.7.5 Equivalent Logarithmic Decrement and Q Factor of Class VI Fractional Vibrators

7.7.5.1 Equivalent Logarithmic Decrement of Class VI Fractional Vibrator

Let t_i and t_{i+1} be two time points of the free response $x_6(t)$, where $x_6(t_i)$ and $x_6(t_{i+1})$ are its successive peak values at t_i and t_{i+1}. Let Δ_{eq6} be the equivalent logarithmic decrement of $x_6(t)$. Then,

$$\Delta_{eq6} = \ln\frac{x_6(t_i)}{x_6(t_{i+1})} = \pi\frac{mw^{\alpha-1}\sin\dfrac{\alpha\pi}{2}+cw^{\beta-1}\sin\dfrac{\beta\pi}{2}+kw^{\lambda-1}\sin\dfrac{\lambda\pi}{2}}{\sqrt{-\left(mw^{\alpha-2}\cos\dfrac{\alpha\pi}{2}+cw^{\beta-2}\cos\dfrac{\beta\pi}{2}\right)kw^{\lambda}\cos\dfrac{\lambda\pi}{2}}\sqrt{1-\dfrac{\left[mw^{\alpha-1}\sin\dfrac{\alpha\pi}{2}+cw^{\beta-1}\sin\dfrac{\beta\pi}{2}+kw^{\lambda-1}\sin\dfrac{\lambda\pi}{2}\right]^2}{4\left[-\left(mw^{\alpha-2}\cos\dfrac{\alpha\pi}{2}+cw^{\beta-2}\cos\dfrac{\beta\pi}{2}\right)kw^{\lambda}\cos\dfrac{\lambda\pi}{2}\right]}}}. \tag{7.236}$$

7.7.5.2 Equivalent Q Factor of Class VI Fractional Vibrator

Denote by Q_{eq6} the equivalent Q factor of a class VI fractional vibrator. Then,

$$Q_{eq6} = \frac{\sqrt{-\left(mw^{\alpha-2}\cos\dfrac{\alpha\pi}{2}+cw^{\beta-2}\cos\dfrac{\beta\pi}{2}\right)kw^{\lambda}\cos\dfrac{\lambda\pi}{2}}}{mw^{\alpha-1}\sin\dfrac{\alpha\pi}{2}+cw^{\beta-1}\sin\dfrac{\beta\pi}{2}+kw^{\lambda-1}\sin\dfrac{\lambda\pi}{2}}. \tag{7.237}$$

7.8 RESULTS FOR CLASS VII FRACTIONAL VIBRATIONS

7.8.1 Equivalent Motion Equation of Class VII Fractional Vibrators

Eq. (7.238) designates the motion equation of a class VII fractional vibrator in the form

$$B_7(t) \triangleq m\frac{d^2x_7(t)}{dt^2}+c\frac{d^\beta x_7(t)}{dt^\beta}+k\frac{d^\lambda x_7(t)}{dt^\lambda}=0. \tag{7.238}$$

The previous equation is equivalently expressed by

$$A_7(t) \triangleq \left(m-cw^{\beta-2}\cos\frac{\beta\pi}{2}\right)\frac{d^2x_7(t)}{dt^2}+\left(cw^{\beta-1}\sin\frac{\beta\pi}{2}+kw^{\lambda-1}\sin\frac{\lambda\pi}{2}\right)\frac{dx_7(t)}{dt}$$
$$+kw^{\lambda}\cos\frac{\lambda\pi}{2}x_7(t)=0. \tag{7.239}$$

As a matter of fact, $F[A_7(t) - B_7(t)] = 0$.

7.8.2 Equivalent Vibration Parameters of Class VII Fractional Vibrators

7.8.2.1 Equivalent Mass of Class VII Fractional Vibrators

Let m_{eq7} be the equivalent mass for a class VII fractional vibrator. From (7.239), we have

$$m_{eq7} = m_{eq2} = m - cw^{\beta-2} \cos \frac{\beta\pi}{2}. \tag{7.240}$$

7.8.2.2 Equivalent Damping of Class VII Fractional Vibrators

Let c_{eq7} be the equivalent damping of class VII fractional vibrators. Then,

$$c_{eq7} = c_{eq6} = cw^{\beta-1} \sin \frac{\beta\pi}{2} + kw^{\lambda-1} \sin \frac{\lambda\pi}{2}. \tag{7.241}$$

In fact, the right side of the previous equation is the coefficient of $\dfrac{dx_7(t)}{dt}$ in (7.239). Thus, (7.241) is true.

7.8.2.3 Equivalent Stiffness of Class VII Fractional Vibrators

Let k_{eq7} be the equivalent stiffness of a class VII fractional vibrator. From (7.239), we have

$$k_{eq7} = k_{eq6} = k_{eq4} = kw^{\lambda} \cos \frac{\lambda\pi}{2}. \tag{7.242}$$

7.8.2.4 Equivalent Damping Ratio of Class VII Fractional Vibrators

Let ζ_{eq7} be the equivalent damping ratio for a class VII fractional vibrator. Define it by

$$\zeta_{eq7} = \frac{c_{eq7}}{2\sqrt{m_{eq7}k_{eq7}}}. \tag{7.243}$$

Substituting m_{eq7}, c_{eq7}, and k_{eq7} into (7.243) yields

$$\zeta_{eq7} = \frac{cw^{\beta-1} \sin \dfrac{\beta\pi}{2} + kw^{\lambda-1} \sin \dfrac{\lambda\pi}{2}}{2\sqrt{\left(m - cw^{\beta-2} \cos \dfrac{\beta\pi}{2}\right) kw^{\lambda} \cos \dfrac{\lambda\pi}{2}}}. \tag{7.244}$$

7.8.2.5 Equivalent Damping-Free Natural Frequency of Class VII Fractional Vibrators

Let ω_{eqn7} be the equivalent damping-free natural angular frequency of a class VII fractional vibrator. Define it by

$$\omega_{eqn7} = \sqrt{\frac{k_{eq7}}{m_{eq7}}}. \tag{7.245}$$

Substituting m_{eq7} and into k_{eq7} (7.245) results in

$$\omega_{eqn7} = \sqrt{\frac{k\omega^{\lambda}\cos\dfrac{\lambda\pi}{2}}{m - c\omega^{\beta-2}\cos\dfrac{\beta\pi}{2}}}. \tag{7.246}$$

7.8.2.6 Equivalent Damped Natural Frequency of Fractional Vibrators of Class VII

Denote by ω_{eqd7} the equivalent damped natural angular frequency for a class VII fractional vibrator. We suppose small damping of $|\zeta_{eq7}| \leq 1$ in what follows from a view of engineering. Define ω_{eqd7} by

$$\omega_{eqd7} = \omega_{eqn7}\sqrt{1 - \zeta_{eq7}^2}. \tag{7.247}$$

Substituting ω_{eqn7} and ζ_{eq7} into (7.247) yields

$$\omega_{eqd7} = \sqrt{\frac{k\omega^{\lambda}\cos\dfrac{\lambda\pi}{2}}{m - c\omega^{\beta-2}\cos\dfrac{\beta\pi}{2}}}\sqrt{1 - \left(\frac{c\omega^{\beta-1}\sin\dfrac{\beta\pi}{2} + k\omega^{\lambda-1}\sin\dfrac{\lambda\pi}{2}}{2\sqrt{\left(m - c\omega^{\beta-2}\cos\dfrac{\beta\pi}{2}\right)k\omega^{\lambda}\cos\dfrac{\lambda\pi}{2}}}\right)^2}. \tag{7.248}$$

7.8.2.7 Equivalent Frequency Ratio of Class VII Fractional Vibrators

Let γ_{eq7} be the equivalent frequency ratio for a class VII fractional vibrator. Define it by

$$\gamma_{eq7} = \frac{\omega}{\omega_{eqn7}}. \tag{7.249}$$

Substituting ω_{eqn7} into (7.249) produces

$$\gamma_{\text{eq7}} = \gamma \sqrt{\frac{1 - 2\varsigma\omega_n\omega^{\beta-2}\cos\dfrac{\beta\pi}{2}}{\omega^\lambda\cos\dfrac{\lambda\pi}{2}}}. \qquad (7.250)$$

7.8.3 Responses of Class VII Vibrators

7.8.3.1 Free Response of Class VII Fractional Vibration Systems

Let $x_7(t)$ be the free response of a class VII fractional vibrator. It is the solution to the following fractional differential equation given by

$$\begin{cases} m\dfrac{d^2x_7(t)}{dt^2} + c\dfrac{d^\beta x_7(t)}{dt^\beta} + k\dfrac{d^\lambda x_7(t)}{dt^\lambda} = 0, \\ x_7(0) = x_{70}, x_7'(0) = v_{70}. \end{cases} \qquad (7.251)$$

Eq. (7.251) can be equivalently expressed by

$$\begin{cases} m_{\text{eq7}}\dfrac{d^2x_7(t)}{dt^2} + c_{\text{eq7}}\dfrac{dx_7(t)}{dt} + k_{\text{eq7}}x_7(t) = 0, \\ x_7(0) = x_{70}, x_7'(0) = v_{70}. \end{cases} \qquad (7.252)$$

From (7.252), therefore, we have

$$x_7(t) = e^{-\varsigma_{\text{eq7}}\omega_{\text{eqn7}}t}\left(x_{70}\cos\omega_{\text{eqd7}}t + \frac{v_{70} + \varsigma_{\text{eq7}}\omega_{\text{eqn7}}x_{70}}{\omega_{\text{eqd7}}}\sin\omega_{\text{eqd7}}t \right), \quad t \geq 0. \ (7.253)$$

7.8.3.2 Impulse Response of Class VII Fractional Vibration Systems

Let $h_7(t)$ be the impulse response of a class VII fractional vibrator. It is the solution to

$$m\frac{d^2h_7(t)}{dt^2} + c\frac{d^\beta h_7(t)}{dt^\beta} + k\frac{d^\lambda h_7(t)}{dt^\lambda} = \delta(t) \qquad (7.254)$$

with zero initial conditions. Since (7.254) can be equivalently expressed by

$$\frac{d^2h_7(t)}{dt^2} + 2\varsigma_{\text{eq7}}\omega_{\text{eqn7}}\frac{dh_7(t)}{dt} + \omega_{\text{eqn7}}^2 h_7(t) = \frac{\delta(t)}{m_{\text{eq7}}}, \qquad (7.255)$$

we have

$$h_7(t) = e^{-\varsigma_{eq7}\omega_{eqn7}t} \frac{1}{m_{eq7}\omega_{eqd7}} \sin\omega_{eqd7}t, \quad t \geq 0. \qquad (7.256)$$

7.8.3.3 Step Response of Class VII Fractional Vibration Systems

Denote by $g_7(t)$ the step response of a class VII fractional vibrator. It is the solution to

$$m\frac{d^2 g_7(t)}{dt^2} + c\frac{d^\beta g_7(t)}{dt^\beta} + k\frac{d^\lambda g_7(t)}{dt^\lambda} = u(t) \qquad (7.257)$$

with zero initial conditions. The equivalent equation of (7.257) is given by

$$\frac{d^2 g_7(t)}{dt^2} + 2\varsigma_{eq7}\omega_{eqn7}\frac{dg_7(t)}{dt} + \omega_{eqn7}^2 g_7(t) = \frac{u(t)}{m_{eq7}}. \qquad (7.258)$$

Thus,

$$g_7(t) = \frac{1}{k_{eq7}}\left[1 - \frac{e^{-\varsigma_{eq7}\omega_{eqn7}t}}{\sqrt{1-\varsigma_{eq7}^2}}\cos\left(\omega_{eqd7}t - \phi_7\right)\right], \quad t \geq 0, \qquad (7.259)$$

where ϕ_7 is given by

$$\phi_7 = \tan^{-1}\frac{\varsigma_{eq7}}{\sqrt{1-\varsigma_{eq7}^2}}. \qquad (7.260)$$

7.8.4 Frequency Transfer Function of Class VII Fractional Vibrators

Let $H_7(\omega)$ be the frequency transfer function of a class VII fractional vibrator. Doing the Fourier transform on both sides of (7.255) yields

$$H_7(\omega)$$

$$= \frac{1}{k\left[\omega^\lambda \cos\frac{\lambda\pi}{2} - \gamma\left(1 - 2\varsigma\omega_n\omega^{\beta-2}\cos\frac{\beta\pi}{2}\right) + i\gamma\left(2\varsigma\omega^{\beta-1}\sin\frac{\beta\pi}{2} + \omega_n\omega^{\lambda-1}\sin\frac{\lambda\pi}{2}\right)\right]}. \qquad (7.261)$$

When writing $H_7(\omega) = |H_7(\omega)|\exp[-\phi_7(\omega)]$, we have

$$|H_7(\omega)| = \frac{1}{k_{eq7}} \frac{1}{\sqrt{\left(1-\gamma_{eq7}^2\right)^2 + \left(2\varsigma_{eq7}\gamma_{eq7}\right)^2}}$$

$$= \frac{1}{k} \frac{1}{\sqrt{\left[\omega^\lambda \cos\frac{\lambda\pi}{2} - \gamma\left(1-2\varsigma\omega_n\omega^{\beta-2}\cos\frac{\beta\pi}{2}\right)\right]^2 + \gamma^2\left(2\varsigma\omega^{\beta-1}\sin\frac{\beta\pi}{2} + \omega_n\omega^{\lambda-1}\sin\frac{\lambda\pi}{2}\right)^2}},$$

(7.262)

and

$$\varphi_7(\omega) = \cos^{-1}\frac{1-\gamma_{eq7}^2}{\sqrt{\left(1-\gamma_{eq7}^2\right)^2 + \left(2\varsigma_{eq7}\gamma_{eq7}\right)^2}}.$$

(7.263)

7.8.5 Equivalent Logarithmic Decrement and Q Factor of Class VII Fractional Vibrators

7.8.5.1 Equivalent Logarithmic Decrement of Class VII Fractional Vibrator

Let t_i and t_{i+1} be two time points of the free response $x_7(t)$, where $x_7(t_i)$ and $x_7(t_{i+1})$ are its successive peak values at t_i and t_{i+1}. Let Δ_{eq7} be the equivalent logarithmic decrement of $x_7(t)$. Then,

$$\Delta_{eq7} = \ln\frac{x_7(t_i)}{x_7(t_{i+1})} = \pi\frac{\frac{c\omega^{\beta-1}\sin\frac{\beta\pi}{2} + k\omega^{\lambda-1}\sin\frac{\lambda\pi}{2}}{\sqrt{\left(m-c\omega^{\beta-2}\cos\frac{\beta\pi}{2}\right)k\omega^\lambda\cos\frac{\lambda\pi}{2}}}}{\sqrt{1-\frac{\left(c\omega^{\beta-1}\sin\frac{\beta\pi}{2} + k\omega^{\lambda-1}\sin\frac{\lambda\pi}{2}\right)^2}{4\left(m-c\omega^{\beta-2}\cos\frac{\beta\pi}{2}\right)k\omega^\lambda\cos\frac{\lambda\pi}{2}}}}.$$

(7.264)

7.8.5.2 Equivalent Q Factor of Class VII Fractional Vibrator

Let Q_{eq7} be the equivalent Q factor of a class VII fractional vibrator. Then,

$$Q_{eq7} = \frac{\sqrt{\left(m-c\omega^{\beta-2}\cos\frac{\beta\pi}{2}\right)k\omega^\lambda\cos\frac{\lambda\pi}{2}}}{c\omega^{\beta-1}\sin\frac{\beta\pi}{2} + k\omega^{\lambda-1}\sin\frac{\lambda\pi}{2}}.$$

(7.265)

7.9 CONCLUDING REMARKS

Principal of seven classes of fractional vibration systems has been concisely surveyed based on Li [51–54]. Due to limited pages of this volume, illustrations regarding fractional vibrations are not described. Refer to Li [51–55] for illustrations. In addition, this chapter has not mentioned stationary sinusoidal responses to seven classes of fractional vibrators, which were recently introduced by Li [55, 56]. Fractional random vibrations addressed in Li [57] will be detailed in Volume II of this monograph. The explanation of nonlinearity of fractional vibrations and the concepts of fractional vibration motions (super-displacement, super-velocity, sub-velocity, super-acceleration, and sub-acceleration) refer to Li [51, Chapter 28].

7.10 EXERCISES

7.1. Prove that $x^{(\alpha)}(t)$ for $1 < \alpha < 2$ or $2 < \alpha < 3$ does not stand for a standard acceleration, where $x(t)$ is a displacement.

7.2. If the unit of m is kg, prove that $mx^{(\alpha)}(t)$ does not designate a standard inertia force for $1 < \alpha < 2$ or $2 < \alpha < 3$, where $x(t)$ is a displacement.

7.3. Prove that $x^{(\beta)}(t)$ for $0 < \beta < 1$ or $1 < \beta < 2$ does not stand for a standard velocity, where $x(t)$ is a displacement.

7.4. Prove that $0 < \lambda < 1$ does not stand for a standard displacement.

7.5. Prove that ζ_{eq1} is not dimensionless when $1 < \alpha < 2$ or $2 < \alpha < 3$.

7.6. Prove that ω_{eqn1} is not a standard damping-free natural angular frequency unless $\alpha = 2$.

7.7. Prove that c_{eq4} takes the form of the Rayleigh damping assumption.

7.8. Write the expression of sinusoidal response to a class I fractional vibrator. (Hint: Li [53])

7.9. Write the expression of sinusoidal response to a class II fractional vibrator. (Hint: Li [53])

7.10. Write the expression of sinusoidal response to a class III fractional vibrator. (Hint: Li [53])

7.11. Write the expression of sinusoidal response to a class IV fractional vibrator. (Hint: Li [54])

7.12. Write the expression of sinusoidal response to a class V fractional vibrator. (Hint: Li [54])

7.13. Write the expression of sinusoidal response to a class VI fractional vibrator. (Hint: Li [54])

7.14. Write the expression of sinusoidal response to a class VII fractional vibrator. (Hint: Li [54])

REFERENCES

1. V. V. Uchaikin, *Fractional Derivatives for Physicists and Engineers*, vol. II, Springer, Berlin, 2013.
2. M. V Shitikova, Impact response of a thin shallow doubly curved linear viscoelastic shell rectangular in plan, *Mathematics and Mechanics of Solids*, vol. 27, no. 9, 2022, 1721–1739.
3. M. V. Shitikova, Fractional operator viscoelastic models in dynamic problems of mechanics of solids: A review, *Mechanics of Solids*, vol. 57, no. 1, 2022, 1–33.
4. Pol D. Spanos and G. Malara, Nonlinear random vibrations of beams with fractional derivative elements, *Journal of Engineering Mechanics*, vol. 140, no. 9, 2014, 04014069.
5. Pol D. Spanos and G. Malara, Nonlinear vibrations of beams and plates with fractional derivative elements subject to combined harmonic and random excitations, *Probabilistic Engineering Mechanics*, vol. 59, 2020, 103043.
6. G. Malara and Pol D. Spanos, Nonlinear random vibrations of plates endowed with fractional derivative elements, *Probabilistic Engineering Mechanics*, vol. 54, 2018, 2–8.
7. J.-S. Duan, The periodic solution of fractional oscillation equation with periodic input, *Advances in Mathematical Physics*, vol. 2013, 2013.
8. J.-S. Duan, A modified fractional derivative and its application to fractional vibration equation, *Applied Mathematics & Information Sciences*, vol. 10, no. 5, 2016, 1863–1869.
9. J.-S. Duan, Z. Wang, Y.-L. Liu, and X. Qiu, Eigenvalue problems for fractional ordinary differential equations, *Chaos, Solitons & Fractals*, vol. 46, 2013, 46–53.
10. J.-S. Duan, Y.-J. Lan, M. Li, A comparative study of responses of fractional oscillator to sinusoidal excitation in the Weyl and Caputo senses, *Fractal and Fractional*, vol. 6, no. 12, 2022, 692.
11. J.-S. Duan, M. Li, Y. Wang, and Y.-L. An, Approximate solution of fractional differential equation by quadratic splines, *Fractal and Fractional*, vol. 6, no. 7, 2022, 369.
12. J.-S. Duan, L. Jing, and M. Li, The mixed boundary value problems and Chebyshev collocation method for Caputo-type fractional ordinary differential equations, *Fractal and Fractional*, vol. 6, no. 3, 2022, 148.

13. J.-S. Duan, D.-C. Hu, M. Li, Comparison of two different analytical forms of response for fractional oscillation equation, *Fractal and Fractional*, vol. 5, no. 4, 2021, 188.

14. M. Zurigat, Solving fractional oscillators using Laplace homotopy analysis method, *Annals of the University of Craiova, Mathematics and Computer Science Series*, vol. 38, no. 4, 2011, 1–11.

15. T. Blaszczyk and M. Ciesielski, Fractional oscillator equation—Transformation into integral equation and numerical solution, *Applied Mathematics and Computation*, vol. 257, 2015, 428–435.

16. T. Blaszczyk, M. Ciesielski, M. Klimek, and J. Leszczynski, Numerical solution of fractional oscillator equation, *Applied Mathematics and Computation*, vol. 218, no. 6, 2011, 2480–2488.

17. T. Blaszczyk, A numerical solution of a fractional oscillator equation in a non-resisting medium with natural boundary conditions, *Romanian Reports in Physics*, vol. 67, no. 2, 2015, 350–358.

18. A. Al-Rabtah, V. S. Ertürk, and S. Momani, Solutions of a fractional oscillator by using differential transform method, *Computers & Mathematics with Applications*, vol. 59, no. 3, 2010, 1356–1362.

19. A. D. Drozdov, Fractional oscillator driven by a Gaussian noise, *Physica A*, vol. 376, no. 2007, 237–245.

20. A. A. Stanislavsky, Fractional oscillator, *Physical Review E*, vol. 70, no. 5, 2004, 051103 (6 pages).

21. B. N. N. Achar, J. W. Hanneken, and T. Clarke, Damping characteristics of a fractional oscillator, *Physica A*, vol. 339, no. 3–4, 2004, 311–319.

22. B. N. N. Achar, J. W. Hanneken, and T. Clarke, Response characteristics of a fractional oscillator, *Physica A*, vol. 309, no. 3–4, 2002, 275–288.

23. B. N. N. Achar, J. W. Hanneken, T. Enck, and T. Clarke, Dynamics of the fractional oscillator, *Physica A*, vol. 297, no. 3–4, 2001, 361–367.

24. A. Tofighi, The intrinsic damping of the fractional oscillator, *Physica A*, vol. 329, no. 1–2, 2003, 29–34.

25. Y. E. Ryabov and A. Puzenko, Damped oscillations in view of the fractional oscillator equation, *Physical Review B*, vol. 66, no. 18, 2002, 184201.

26. M. S. Tavazoei, Reduction of oscillations via fractional order pre-filtering, *Signal Processing*, vol. 107, 2015, 407–414.

27. T. Sandev and Z. Tomovski, The general time fractional wave equation for a vibrating string, *J. Physics A: Mathematical and Theoretical*, vol. 43, no. 5, 2010, 055204 (12pp).

28. H. Singh, H. M. Srivastava, and D. Kumar, A reliable numerical algorithm for the fractional vibration equation, *Chaos, Solitons & Fractals*, vol. 103, 2017, 131–138.

29. L.-F. Lin, C. Chen, S.-C. Zhong, and H.-Q. Wang, Stochastic resonance in a fractional oscillator with random mass and random frequency, *Journal of Statistical Physics*, vol. 160, no. 2, 2015, 497–511.

30. L.-F. Lin, C. Chen, and H.-Q. Wang, Trichotomous noise induced stochastic resonance in a fractional oscillator with random damping and random frequency, *Journal of Statistical Mechanics: Theory and Experiment*, vol. 2016, 2016, 023201.

31. H. S. Alkhaldi, I. M. Abu-Alshaikh, and A. N. Al-Rabadi, Vibration control of fractionally-damped beam subjected to a moving vehicle and attached to fractionally-damped multi-absorbers, *Advances in Mathematical Physics*, vol. 2013, 2013.

32. H. Dai, Z. Zheng, and W. Wang, On generalized fractional vibration equation, *Chaos, Solitons & Fractals*, vol. 95, 2017, 48–51.

33. R. Ren, M. Luo, and K. Deng, Stochastic resonance in a fractional oscillator driven by multiplicative quadratic noise, *Journal of Statistical Mechanics: Theory and Experiment*, vol. 2017, 2017, 023210.

34. R. Ren, M. Luo, and K. Deng, Stochastic resonance in a fractional oscillator subjected to multiplicative trichotomous noise, *Nonlinear Dynamics*, vol. 90, no. 1, 2017, 379–390.

35. Y. Xu, Y. Li, D. Liu, W. Jia, and H. Huang, Responses of Duffing oscillator with fractional damping and random phase, *Nonlinear Dynamics*, vol. 74, no. 3, 2013, 745–753.

36. G. He, Y. Tian, and Y. Wang, Stochastic resonance in a fractional oscillator with random damping strength and random spring stiffness, *Journal of Statistical Mechanics: Theory and Experiment*, vol. 2013, 2013, P09026.

37. A. Y. T. Leung, Z. Guo, and H. X. Yang, Fractional derivative and time delay damper characteristics in Duffing-van der Pol oscillators, *Communications in Nonlinear Science and Numerical Simulation*, vol. 18, no. 10, 2013, 2900–2915.

38. L. C. Chen, Q. Q. Zhuang, and W. Q. Zhu, Response of SDOF nonlinear oscillators with lightly fractional derivative damping under real noise excitations, *The European Physical Journal Special Topics*, 193, vol. no. 1, 2011, 81–92.

39. J.-F. Deü and D. Matignon, Simulation of fractionally damped mechanical systems by means of a Newmark-diffusive scheme, *Computers & Mathematics with Applications*, vol. 59, no. 5, 2010, 1745–1753.

40. G. E. Drăgănescu, L. Bereteu, A. Ercuţa, and G. Luca, Anharmonic vibrations of a nano-sized oscillator with fractional damping, *Communications in Nonlinear Science and Numerical Simulation*, vol. 15, no. 4, 2010, 922–926.

41. Y. A. Rossikhin and M. V. Shitikova, New approach for the analysis of damped vibrations of fractional oscillators, *Shock and Vibration*, vol. 16, no. 4, 2009, 365–387.

42. F. Xie and X. Lin, Asymptotic solution of the van der Pol oscillator with small fractional damping, *Physica Scripta*, vol. 2009, No. T136, 2009, 014033.

43. J. Yuan, Y, Zhang, J, Liu, B, Shi, M. Gai, and S. Yang, Mechanical energy and equivalent differential equations of motion for single-degree-of-freedom fractional oscillators, *Journal of Sound and Vibration*, vol. 397, 2017, 192–203.

44. Y. Naranjani, Y. Sardahi, Y.-Q. Chen, and J.-Q. Sun, Multi-objective optimization of distributed-order fractional damping, *Communications in Nonlinear Science and Numerical Simulation*, vol. 24, no. 1–3, 2015, 159–168.

45. A. Di Matteo, P. D. Spanos, and A. Pirrotta, Approximate survival probability determination of hysteretic systems with fractional derivative elements, *Probabilistic Engineering Mechanics*, vol. 54, 2018, 138–146.

46. Ž. Tomovski and T. Sandev, Effects of a fractional friction with power-law memory kernel on string vibrations, *Computers & Mathematics with Applications*, vol. 62, no. 3, 2011, 1554–1561.

47. J. F. Gomez-Aguilar, J. J. Rosales-Garcia, J. J. Bernal-Alvarado, T. Cordova-Fraga, and R. Guzman-Cabrera, Fractional mechanical oscillators, *Revista Mexicana de Fisica*, vol. 58, no. 4, 2012, 348–352.

48. Y. Tian, L.-F. Zhong, G.-T. He, T. Yu, M.-K. Luo, and H. E. Stanley, The resonant behavior in the oscillator with double fractional-order damping under the action of nonlinear multiplicative noise, *Physica A*, vol. 490, 2018, 845–856.

49. M. Berman and L. S. Cederbaum, Fractional driven-damped oscillator and its general closed form exact solution, *Physica A*, vol. 505, 2018, 744–762.

50. A. Coronel-Escamilla, J. F. Gómez-Aguilar, D. Baleanu, T. Córdova-Fraga, R. F. Escobar-Jiménez, V. H. Olivares-Peregrino, and M. M. Al Qurashi, Bateman-Feshbach Tikochinsky and Caldirola-Kanai oscillators with new fractional differentiation, *Entropy*, vol. 19, no. 2, 2017 (55 pages).

51. M. Li, *Fractional Vibrations with Applications to Euler-Bernoulli Beams*, CRC Press, Boca Raton, 2023.

52. M. Li, *Theory of Fractional Engineering Vibrations*, Walter de Gruyter, Berlin/Boston, 2021.

53. M. Li, Three classes of fractional oscillators, *Symmetry-Basel*, vol. 10, no. 2, 2018 (91 pages).

54. M. Li, Analytic theory of seven classes of fractional vibrations based on elementary functions: A tutorial review, *Symmetry*, vol. 16, no. 9, 2024, 1202.

55. M. Li, Stationary responses of seven classes of fractional vibrations driven by sinusoidal force, *Fractal and Fractional*, vol. 8, no. 8, 2024, 479.

56. M. Li, Dealing with stationary sinusoidal responses of seven types of multi-fractional vibrators using multi-fractional phasor, *Symmetry*, vol. 16, no. 9, 2024, 1197.

57. M. Li, PSD and cross PSD of responses of seven classes of fractional vibrations driven by fGn, fBm, fractional OU process, and von Kármán process, *Symmetry*, vol. 16, no. 5, 2024, 635.

CHAPTER 8

Postscript to Volume I

I N CHAPTER 5, BASED ON several cases of structures with frequency-dependent elements, I brought forward a general form of fractional vibration systems with frequency-dependent elements (mass, damping, and stiffness) and its analytic expressions of equivalent motion equation, equivalent elements (mass, damping, and stiffness), equivalent vibration parameters (damping ratio, natural angular frequencies, frequency ratio), responses (free, impulse, and step), equivalent frequency transfer function, equivalent logarithmic decrement, and equivalent Q factor. Chapter 5 does not relate to the knowledge of fractional calculus and fractional vibrations accordingly. However, the general form of fractional vibration systems with frequency-dependent elements in Chapter 5 may be taken as a pre-model of fractional vibration systems discussed in Chapters 6 and 7. As a matter of fact, my intention in writing Chapter 5 is to attract readers' attention to the fractional vibration systems in Chapters 6 and 7 because they are with frequency-dependent elements. For instance, a class I fractional vibration system is with frequency-dependent mass and damping. So are class II and class III fractional vibration systems. A class IV fractional vibrator is with frequency-dependent mass, damping, and stiffness. So are class VI and class VII fractional vibrators. A class V fractional vibrator is only with frequency-dependent damping and stiffness.

For seven classes of fractional vibration systems, in Chapter 7, we explained the closed forms of their equivalent motion equations, equivalent elements (mass, damping, stiffness), equivalent vibration parameters (damping ratio, natural angular frequencies, frequency ratio), responses (free, impulse, and step), and frequency transfer functions using

DOI: 10.1201/9781003657897-8

elementary functions. Each class may be taken as a special model of the general form of vibration system with frequency-dependent elements discussed in Chapter 5.

Volume I is taken as a theory of fractional random vibrations. From the point of view of engineering, they are not enough. Two theoretic directions below may be worth investigation in future.

- Level crossing processes of fractional processes.

- Crack issues and or fatigue issues associated with fractional vibration systems excited by fractional processes.

If one feels that this volume may be a pavement with respect to fractional random vibrations driven by fractional processes, I will be honoured to be a paver in this regard.

Index